JN117189

安定化原理に基づく エネルギー発生

赤沼 篤夫

Energy Development on the Stabilization Principle

Akanuma Atsuo

東京図書出版

Contents ── 目次

Energy Development on the Stabilization Principle

Prologue

The existence of actual point which is represented by a mass point is based on the matching with its environments, which is the stability. The theory on actual points which change to match their existence environment is summarized as the theory of stability. Actual points or mass points have not only stable existence but also movements which cause variations. These movements bring forth stability, which is stabilization. The stabilization means the elevation of existence probability, which is an acceptable principle. This principle explains the generation of energy. The fact was pursued here that the existence is the stabilized energy.

Contents

A. Space and existence

1. Conceptual and material existences

Conceptual existences which we think of are stable and perceived in human mind with scholar values. They are subjective. But real natural existences are objective and are essentially independent from the human concepts. Natural actualities are mostly stable on the human thought. The most of spaces are stable in human mind and consist of three dimensions with infinite number of conceptual points which are subjective and have neither dimensions nor instabilities. Spaces have stable three dimensions and contain conceptually stable existents. The space conceptually consists of infinite number of these conceptual points.

But space has time which disturbs the stability and gives objective variability to these stable existents which gain objective instability. Those points with physical actuality abide with the natural regulations which are essentially contingent. The existences of those actual or objective points are objective products from the contingency and are controlled by the regulations induced from the contingency. They are denominated objective existences and follow the natural regulations which are essentially contingent. Contingency comprises stability and instability. Contingency can make stable existents lose balance with surroundings then they become unstable existents. This balance may be natural precarious factors.

2. Subjective and objective points

Point has been defined by Euclid. It has neither length nor space. It has neither volume nor substantiality. It can simply show exact position. These points have no substantiality and indicate the conceptual positions. They also indicate their subjective relations among them. Those points involve little in physical events except indicating positions.

Some points are involved in physical events. They have objective unstable existence and movement. An example is a mass point which represents the substantial objective existence and movement although it involves some instability. A point which involves any physical events becomes a substantial point.

3. Subjective and objective spaces

Euclid defined the space is three dimensional and consists of infinite number of Euclid points. The space in our vicinity is three dimensional and is considered consisting of infinite number of ontologically defined points which are conceptual. The great space of universe exists but it has not been perfectly understood. It is usually considered three dimensional. Its nature has to be still much investigated. Various multi-dimensional spaces are employed in various studies. They are essentially conceptual. The stable three-dimensional space is conceptual and is still subjective.

The space of our vicinity has been perceptible as the three-dimension with time which affords the objective spatial and time-dependent instabilities. Objective space has position and motion which make some instability on reference to the primary space which can be conceptual subjective or virtual objective space. Hence the objective space carries some precarious agents which make the space somewhat unstable. Hence the entire points in objective space have the instability of existence and persistence, which allow substantiality to entire objective space. Objective space is substantial existence. Objective spaces consist of large number of objective points and infinite time. The current objective space of our vicinity is the space of temperature, which is attributable to the constituent objective points in the objective space of the universe.

4. Subspace

Space can have another secondary space in it. The reference frame for the secondary space is the primary space. The primary space for the reference frame is an objective space or can be subjective space. Hence the objective secondary space has some instability of position and motion and has in consequence substantiality. The subspace which is substantial space has always three dimensions with its own time. Time makes instability. An objective point existing at a position of a subspace coordinate system has some instability. An objective point stays at one position since this position is the most probable position for it, although this position implies some improbability, therefore, it has some instability of existence and persistence.

B. Substantiality

1. Time

The substantiality of existences has been often considered without the factor of time which is the most important factor. The existence of objects should be considered with time in the three-dimension space. Hence the concept of time is required. The time considered in a space is elapsed time. The elapsed time which is called here time is the duration from one moment to another in a specific space. Substantial spaces have always their own time. Several other secondary spaces are considered to exist in one primary space. Each substantial secondary space has its own time whose value is also statistic. In objective spaces the relativity of time occurs on dynamic events. Time is an independent variable existing in a space. Distance in a fundamental direction, for example, can become dependent variable of time. Space and time construct a frame of existence. Space and time are required to express the existence of a physical event. The existence of an objective point can be expressed in a space-time frame which is the substantial space the objective point belongs to.

2. Conceptual and substantial points

The existence of events has been considered in a three-dimension space without time which is named Euclidean space and is stable conceptual space. A point shows constant position and has no dimension which is Euclid point. Its existence is conceptual and is designated here a conceptual or subjective point. A Euclid point exists stable at one position of a coordinate system and its existence is stable.

But substantial or objective points have some instability. They have insecurity and inconstancy and are designated here substantial or objective points. An example is the gravity center point of a ball in a space. It is not stable and has insecurity and inconstancy. It is an objective point which has positional insecurity and inconstancy. A mass point is an example of objective point which has substantiality. Mass point without mass is the substantial point which is named here objective point. Objective point is actuality and has various value of substantiality. The objective point which represents the existence of physical events has position and motion with existence and persistence. To express its steadiness each space has to have its own time axis. The existence of events must be recognized with time. The space where objective points reside should be determined with a set of an origin point and three fundamental directions together with time. An objective point has position and motion which consist of velocity and acceleration with respect to time.

3. Substantial space

To express the position and motion of an objective point a scale or a set of coordinates is employed. It represents a stable space where the point exists with the stable condition. This conceptual space is determined with one most fundamental point which is the origin and three fundamental directions. The most fundamental space is conceptual and has infinite number of Euclid points.

The most fundamental objective point can be collated with a point of a conceptual space and forms a subspace which is an objective space with relativity to the original conceptual space. There should be objective substantial subspace which consists of points with insecurity and inconstancy.

Substantial or objective space with large number of objective points has also existence and persistence probability distributions and is unstable since the origin point of this objective space is an objective point. Space carries various existents of events with substantiality including objective points. Contingency can happen to bring forth insecurity and inconstancy to those existences, therefore, they carry variability and continuity and in consequence substantiality.

An objective space has its own time and coordinates system. A space which has one basic point and three fundamental directions gives secondary spaces which are movable, therefore, they are unstable. Space has subspaces in it.

The minimum size of subspace is not zero size since the zero size is a point. All points in subspaces have the existence provability with insecurity and the persistence provability with inconstancy. Hence they have variability and continuity. And they have in consequence substantiality.

A micro-space around a point in an objective space has substantiality. This point represents this micro-space which is named cubage. The cubage should be regarded as this point which is an objective substantial point. This point represents the micro-space of which size is not determined.

C. Stabilization principle

Objective points get stabilized according to the statistical probability, which is the stabilization principle. Stabilization means the elevation of statistical probabilities. The local maximum statistical probability brings forth the stability. Position and motion are ought to be handled with not scholar values but statistical values. A simple example is measurement of one meter. This one meter is conceptual value. It is actually the most provable value which has small error possibility.

Events change toward the more probable state. Existence and persistence of events follow the reliabilities of their provability; therefore, their values have to be expressed statistically. Events vary according to their provability. When the reliability of provability goes up its potentiality goes up which is negative entropy, therefore, its entropy goes down. Potentiality is essentially negative entropy. When stability goes up some negative entropy must work on it.

As a simple example an event follows. A glass of cold 0 degrees water is left in air space of which temperature is 25 degrees. The water warms up to 25 degrees in time, which means the stabilization. Temperature difference is potential difference which disappears in time decreasing the entropy of cold water, which is the conceptual stabilization with scholar values and follows. Starting temperature is 0 degree and final temperature is 25 degrees. In the conceptual resolution the potential increase of water is proportional to temperature difference as is shown in equation below where x is temperature, t is time, a is the most reliable temperature and b is conversion rate.

$$bx = \mathrm{d}(a-x)/\mathrm{d}t$$

This equation can be changed as follows.

$$\mathrm{d}x/x = -b^{*}\mathrm{d}t$$

The solution of this equation which is stabilization curve is following abducting function.

$$x = a^*[1 - \exp^*(-bt)]$$

a is the most reliable water temperature

b is conversion rate

These equations are conceptual. This x has scholar value. These equations express the water temperature stabilization. The stabilization principle induces this equation applying its variability which is a proposition of stabilization principle where x is statistic value which has existence and persistence. The existence of above example is expressed as follows.

Equation (1) shows the existence situation.

$$(1) \qquad \Phi = \exp[-(x-a)^2/(2\sigma^2)]/[\sigma(2\pi)^{1/2}]$$

σ is standard deviation of Φ $\quad \sigma > 0$

x is temperature of water in the glass which will be changed by air temperature and a is the most reliable water temperature in this situation.

For convenience X shall be deviation temperature from the expected most reliable temperature (a), then $X = x - a$. Equation (1) becomes as follows.

X is the existence from expected point and V is its persistence which is dx/dt. They are statistic states.

X shall be deviation on axis x from the expected point (a) which is $X = x - a$.

The situation of this event is expressed by the following two equations which are processed and will stabilize.

$$(2) \qquad \Phi = \exp[-X^2/(2\sigma^2)]/[\sigma(2\pi)^{1/2}]$$

$$(3) \qquad \Psi = \exp[-V^2/(2\tau^2)]/[\tau(2\pi)^{1/2}]$$

These equations are applied to the variability which is Φ*Ψ and results to the equation $\mathrm{d}x/\mathrm{d}t = -bx$. This equation is the same as the one employed in the conceptual process shown above. The process will be shown later in the variability section.

Existence and its persistence are statistically determined not with scholar values. Any event of objective point has existence and persistence which carry probability distributions and varies according to their probability. Objective points stabilize increasing their reliability. The situation of high reliability means unstrained condition. Actual existence implies some insecurity. An objective point which is objective existence and is represented by a point in space is always affected by time, therefore, it is always accompanied by persistence. Objective presence implies some inconstancy. Space makes existence and time makes persistence. In above example existence is temperature x, persistence V which is $\mathrm{d}x/\mathrm{d}t$ and is usually expressed with V.

The stabilization principle consists of following two constituents of existence and persistence which are axioms.

1. Insecurity; Events exist. Existence involves some statistical insecurity.
2. Inconstancy; Events persist. Persistence involves some statistical inconstancy.

D. Existence and persistence of objective points

1. Existential insecurity and inconstancy

(1) Existential uncertainty

A mass point which is an existent has position and motion. So does an objective point without mass. The objective point has position and motion. Its position has some uncertainty which is insecurity. And its motion has some unsteadiness which is inconstancy. A point which a man thinks about is a conceptual point. It can be placed exactly at a position in a conceptual space and can make free movements. The point stays at a position in a conceptual space with perfect accuracy and its movement has complete steadiness.

An objective point based on nature can be considered as the simplest existence. It does not stay or move exactly as a man thinks about. It is an existent which does not concern on human thought and follows natural laws. Objective existent which follows natural laws is involved with contingency. A point on an object exists with certain insecurity and is independent from human thought. In a physical space there is essential difference between the existence of the conceptual point and the existence of an objective point which has shears and shakes. An objective point has position and motion. Physical existence of an objective point has positional insecurity and inconstancy. The position can comprise gaps and deviations, which are designated insecurity. And motion can have fluctuation, which is designated inconstancy.

(2) Stabilization of events

Events exist and alter. The existence of an objective point is an event.

Every event has primary quantity and secondary quantity with at least one communal parameter which causes statistical instabilities of both quantities. About electric quantities, for example, the electric charge is primary quantity which is determined by voltage but can alter. The alteration of this charge is electric current which is secondary quantity and is determined by voltage. Communal parameter is voltage. The secondary quantity is the alteration of the primary quantity. Another example is a mass point. A mass point resides at one point, which is primary quantity and is determined by distance. This existence has possibility to alter from the expected position, which is movement. This movement is secondary quantity and determined by quantities including distance. Distance is the communal parameter and causes the instability of existence and movement of a mass point. This logic is applicable to objective points without mass. There should be deviations from the expected existence point. The deviation is statistic. The existence reliability changes according to deviations. Every existence is hardly perfect. The reliability of an existence with little deviation may be high but not perfect. The existential probability at small deviation is comparatively high. The probability distribution of the existence is normal distribution centering no deviation. Actual stable existence of an objective point at one position means probably stabilized presence at the position with existence and persistence. It is the most probable presence, which implies a little improbability. Physically existing points have persistence and follow probability. An objective point at a position stays there since this position is the most probable position, although this position implies a little improbability, therefore, it has some instability of existence and persistence.

The nature is actualized by the stabilization which the contingency brings forth. Events including the existence of objective points follow the statistical stabilization processes. The statistical perception on the essentialities of natural events induces this stabilization. Events vary according to their probability, which is designated here the stabilization principle.

2. Existence

(1) Objective existence

An objective point in a space is considered to have position and motion. The position of a point has relativity to another point and the motion is the change of this relativity. The position of a point changes then its relativity, for example, to the origin changes. The above statement is the conceptual perception of existence. This concept of position and motion is employed in the perception of objective point locations which should have in nature some instability. When the point is not conceptual but substantial it does not always coincide perfectly to the corresponding point of the reference frame. It can shear from the corresponding point of the reference frame. The coincidence is conceptual and it is actually not perfect. Its exactitude is statistical. The coincidence is not secured. Shear or deviation accompanies to the coincidence. Shear or deviation decreases the statistical exactitude, which is existential insecurity. The expected position which is usually the center of existential probability distribution has the higher reliability. When a objective point would not be in the center it would move to the position with higher reliability which resides usually at the center. The reliability difference causes the centripetal force, which is stabilization. The space defines the existence and the time defines the persistence in space. The physical objects in space have position and motion collating to the coordinate system of existing space, which induces dispersion and fluctuation and, therefore, they have insecurity and inconstancy. Existence varies because of time. Space makes insecurity. Time makes inconstancy.

A solid object, which is solidified energy or substance, is also considered as an aggregation of objective points. Therefore, its presence can be expressed in a reference frame. The presence of a substantial object in a substantial space does not necessarily coincide to the human conceptual presence. The human thought is a scale and its quantity is scholar. The substantial quantity measured with this scale is statistic. An aggregation of points with insecurity in this meaning is a physical object or substance.

A substantial existence is often represented by one geometrical point neglecting its shape and size. This geometrical point represents the existence, which is the objective point. Presence of energy is often represented by one geometrical point and the physical quantity of energy is given to this point neglecting its shape and size.

(2) Existential insecurity

Events exist and continue or vary. Existence is statistic quantity and involves some statistical instability, which is insecurity. The reliability in the existential probability distribution of the objective point is the maximum at no dispersion or at the expected location and around it there lie lower probability densities. This situation is involved with high or low precision of the existence and is designated insecurity.

This situation is easily conceivable with particle locations. When the position of electrons or other particles at one instance is considered, the position can be calculated. But not all particles are at the calculated position. The value comprises some statistical dispersion which is insecurity. The calculated position is the modal value. Some existences are scattered around it. It is the statistical existence.

(3) Existence probability distribution

Existential precision is not perfect. Existence always comprises some insecurity which is expressed with standard deviation (σ). Existential probability distribution indicates the possible existence at the expected point and its surroundings. The deviation from the expected point is the stochastic variable. The expected position is the modal value. It is the mean value of the variable in the distribution which is a normal distribution. The reliability at the modal value is the existential precision. The higher value of this reliability is the more stable is the existence. The value of standard deviation defines insecurity. The smaller value of the standard deviation has the more stable the existence is. Its precision is defined by the reciprocal of standard deviation. To make it simple, the distribution of the x direction is considered. The occurrence on axis x is all over the same. Hence the probability distribution (Φ) is the normal distribution with a standard deviation (σ) and the expected point which is ($x = a$) in the center.

An example of existence which is shown in section C shall be expressed as follows and processed with the variability. The situation of this existence is expressed by the following two equations which will stabilize. Following equation shows the existence situation.

$$(1) \qquad \Phi = \exp[-(x-a)^2/(2\sigma^2)]/[\sigma(2\pi)^{1/2}]$$

σ is standard deviation of Φ $\quad \sigma > 0$

x is temperature of water in the glass which will be changed by air temperature and a is the most reliable water temperature of this existence.

X shall be deviation temperature from the expected most reliable temperature (a), then $X = x-a$. Equation (1) becomes as follows.

$$(2) \qquad \Phi = \exp[-X^2/(2\sigma^2)]/[\sigma(2\pi)^{1/2}]$$

Reliability of the expected temperature ($X = 0$) is $1/[\sigma(2\pi)^{1/2}]$. The higher this value is the steeper decreases or increases the reliability, therefore, small gaps from (a) exist with high reliability, which is precise existence of temperature. Reliability decreases as standard deviation (σ) increases, therefore, the magnitude of (σ) represents insecurity and its reciprocal ($1/\sigma$) expresses existential precision.

(4) Occurrence and insecurity

Let E be the energy at the expected point and let Pc be the occurrence of a unit of energy. Then the relationship between standard deviation (σ) of existence probability distribution and the occurrence (Pc) of unit of energy is the following,

(4) $$\sigma^2 = Pc(1-Pc)/E$$

Pc is small enough compared to 1

(5) $$\sigma^2 \fallingdotseq Pc/E$$

The standard deviation is the square root of the quotient of occurrence and energy. Mean variance is the same to the occurrence of a unit of energy or the occurrence rate. Hence the square root of energy and the reliability of existence probability are proportional. And the reliability and the square root of occurrence are inversely proportional. An objective point in the space of uniform occurrence has all the same mean variance of existence probability at any point in the space. An objective point in that space is stable and has no motion. But the reliability of existence probability is not 100%. The possibility of changing its position coexists.

E. Persistence and inconstancy

1. Inconstancy

Among the existences of the same precision, some objective points easily change their positions and some points hardly change. When the existence at a position is considered it has not only displacement possibility but also fluctuation possibility. An existence has essentially two elements of instability which are existential insecurity and inconstancy. Existence has not only positional instability whether a point is on the expected position, which is insecurity, but also qualitative instability of persistent motion. The latter instability is designated here inconstancy. Steady motion with high persistence has little motion fluctuation. There are existences which are easy to fluctuate with high inconstancy or those which are hard to fluctuate with low inconstancy. This difference is expressed by the difference of inconstancy. The low inconstancy means that a point has strong fixture to the position, which is high persistence. Existence generally implies the two elements of precision and consistency, which complies that an objective point on a position has not only deviation but also fluctuation.

2. Persistence

Events persist. Persistence is statistic quantity and involves some statistical instability, which is inconstancy. In every event the existence persists or alters. The alteration is the secondary quantity. The alteration of existence is also statistic. The probability of rapid alteration should be low. Probability of low rapidity is comparative high. Probability of no alteration which is persistence is high but not 100%. The persistence probability distribution is normal distribution centering no alteration.

Events have two major factors which are existence and persistence. About the example of mass point the existence of a mass point has position and motion. Motion is the persistence of position, which is obtained differentiating the distance with a parameter of time. Position has the existential probability distribution and motion has persistence probability distribution. Existence has insecurity and persistence has inconstancy. These two quantities are statistic. Events have two major factors which are existence and persistence and have statistic values of insecurity and inconstancy.

An objective point has two elements which are position and motion. Motion instability distributes with normal distribution centering no fluctuation. Its standard deviation expresses the magnitude of the inconstancy. The stochastic variable of this persistency probability distribution is the time dependent change rate of the stochastic variable of existence probability distribution. This change rate is designated here fluctuation velocity or fluctuation. A standstill objective point has fluctuation modal value of 0 and its reliability is the standstill possibility. The probability distribution is a normal distribution and the smaller is its standard deviation the higher is the standstill possibility and the more persistent. Hence the reciprocal of the standard deviation is designated here persistency. When a objective point is standstill at a point (a) in a space then the fluctuation velocity is 0. There is possible fluctuation even in the standstill situation.

3. Persistence probability distribution

Let X be the deviation of an objective point staying at position (a) on axis x, then the fluctuation velocity (V) is as follows.

(6) $$\mathrm{V} = \mathrm{dX}/\mathrm{d}t$$

Fluctuation is the time dependent change rate of deviation and has the dimension of velocity. An objective point has fluctuation probability of normal distribution centering 0 on any direction. Let (Ψ) be the persistence probability distribution, then its standard deviation (τ) expresses the inconstancy and its reciprocal is the consistency or strength of persistence. Considering on axis x, an objective point staying still at ($x = a$) has a normal distribution of inconstancy centering 0 of fluctuation velocity.

$$\Psi = \exp[-V^2/(2\tau^2)]/[\tau(2\pi)^{1/2}] \quad (3)$$

Reliability at $V = 0$ is $1/[\tau(2\pi)^{1/2}]$, which is the steadfast reliability. The higher this value is, the more makes abrupt fall of reliability with a small shift.

A point at deviation (X) has the persistence probability which is expressed with normal distribution (Ψ) as equation (3). Persistence is not perfect. Persistence always comprises some inconstancy which is expressed with standard deviation (τ).

F. Variability and continuity

1. Stability of existence

Existence is accompanied with persistence. They form variability continuity and substantiality. Events including the existence of objective points have variability and continuity. When the existing situation changes, events vary for stabilization and stabilize for continuity. Events have variability continuity and in consequence substantiality. On variability (Z) the probabilities of existence (Φ) and persistence (Ψ) are proportional and on continuity the probabilities of existence and persistence are inversely proportional. Existence of objective point is accompanied with persistence. They form variability, continuity and they substantialize events. Objective points have to have substantiality. Variability and continuity are anti-proportional in substantiality. Hence substantiality disappears when either insecurity or inconstancy becomes zero. When an existent loses substantiality it disappears in space. Existents must have substantiality. As discussed later when the reliabilities of existence and persistence are at the maximum inconstancy and insecurity are proportional in variability as equation (7) shows and inconstancy and insecurity are anti-proportional in continuity (8). Variability and continuity are anti-proportional in substantiality (9). Then either insecurity or inconstancy becomes zero substantiality disappears. When an existent loses substantiality it disappears in space. Events exist in the balance of variability and continuity, which is a stabilized property of substantiality.

2. Variability of an objective point

In variability insecurity and inconstancy are in the same direction. Insecurity and inconstancy are proportional in variability. Variability is defined as the quotient of persistence probability and existence probability. As described later the variability (Z) is a non-time dependent constant which is independent to the elapse of time.

A conceptual point can move free in a conceptual space. But substantial points cannot move freely in space. The motion is time dependent. Besides the existence probability restricts the motions. Particularly the relativity change of each point on a shape to the origin of a substantial space is restricted. Substantial or actual motion depends not only on persistence probability but also existence probability. Considering on motion the concept of duration time is required. The elapsed time which is called here time is the duration from one moment to another. Motion changes the existence reliability and stabilizes the objective point, which is variability. The variability comprises the elapse of time. An objective point with high inconstancy is apt to change and together with high insecurity it becomes more changeable.

For variability insecurity and inconstancy are in the same direction. Insecurity and inconstancy are proportional in variability. Variability is defined as the quotient of existence probability and persistence probability as shown on equation (7). As described later the variability (Z) is a constant which is independent to the elapse of time.

$$(7) \qquad Z = \Phi/\Psi$$

Logarithm of both sides equation (7) are taken and differentiated,

$$(8) \qquad (d\Phi/dt)/\Phi-(d\Psi/dt)/\Psi = (dZ/dt)/Z$$

The variability induces that the remainder of existence variation rate and persistence variation rate is variability variation rate. Equations (2) and (3) are substituted into equation above then equation below is obtained.

$$(9) \qquad (dX/dt)X/\sigma^2-(dV/dt)V/\tau^2 = (dZ/dt)/Z$$

X is deviation from air temperature (a) and V = dX/dt. Z is a constant.

Then equation below is obtained.

(10) $d^2X/dt^2/\tau^2 - X/\sigma^2 = 0$ Let $\tau/\sigma = \omega$ then

(11) $d^2X/dt^2 - \omega^2 X = 0$

above equation (11) can be changed as follows.

(12) $[d/dt+(\tau/\sigma)][d/dt-(\tau/\sigma)]X = 0$

(13) $dX/dt-(\tau/\sigma)X = 0$ or

(14) $dX/dt+(\tau/\sigma)X = 0$

Since equation (13) diverges to infinite. It is a destructive process and is not realistic. This is a stabilizing process and here ($dX/dt+(\tau/\sigma)X = 0$) should be applied. The solution of equation (12) shall be as follows.

(14) $dX/dt+(\tau/\sigma)X = 0$

This equation is essentially the same to $dx/x = -b*dt$ which is discussed in the clause of Stability where $\tau/\sigma = a$, $X = x-a$.

The solution of above equation is following abducting function.

(15) $X = X_0*\exp[-(\tau/\sigma)t]+C$

C is a constant

Actually continuity works on this phenomenon and this curve becomes an abducting waving curve.

3. Continuity of an objective point

For continuity events with high insecurity must have low inconstancy and events with high inconstancy must have low insecurity.

The substantial stability consists of variability and continuity. Natural events carry on or change. Changes are essentially for stability which is continuity. When situations change, events vary for continuity. For continuity events with high insecurity must have low inconstancy and events with high inconstancy must have low insecurity. In continuity of events insecurity and inconstancy are in opposite direction. Hence insecurity and inconstancy are inversely proportional in continuity. Persistence reliability and existence reliability are inversely proportional. Continuity is defined as the product of existence probability and persistence probability as shown with equation (16). As described later the continuity (H) is a constant factor.

$$(16) \qquad H = \Phi * \Psi$$

4. Substantiality

When an event is in changing situation its substantiality has large variability and its continuity should be small. And when it is in stabilized condition its variability is small and its continuity is large. Variability and continuity are in opposite direction. They are inversely proportional in substantiality of physical events. Events with low continuity change easy having high variability but in situations with low variability they have high continuity. Stabilized events with high continuity have low variability. Hence the substantiality (Θ) is defined as the product of variability (Z) and continuity (H) as is shown with equation (17). Substantiality (Θ) is a constant and expresses the characteristic of an event and its essentiality.

$$(17) \qquad \Theta = Z * H$$

On the substantiality (Θ) of an event variability (Z) and continuity (H) are inversely proportional. It is expressed with above equation (17) which clarifies that variability (Z) and continuity (H) are constants.

Substantiality is constant since it expresses the characteristics of existents or events. Then variability and continuity are also constants under constant conditions.

Logarithm of both sides of equation (17) is taken,

$$\log(Z)+\log(H) = \log(\Theta)$$

And both sides are differentiated. Θ is a constant since substantiality is the characteristics of events.

Hence $d\Theta/dt/\Theta$ is 0.

$$dZ/dt/Z+dH/dt/H = 0 \quad (18)$$

When the surrounding conditions do not change its continuity (H) does not change. Hence $dH/dt/H$ is 0. In consequence $dZ/dt/Z$ is also 0. H and Z are constants under constant conditions.

Real beings have continuity together with variability which is caused by the existence factors of insecurity and inconstancy. The variability may induce new substantiality which is new existence and has new continuity and variability.

G. Revolution and stabilization

1. Revolution of events

(1) Variability and revolution

Existence reliability and persistence reliability are proportional in variability. Hence the variability is the quotient of existence probability (Φ) and persistence probability (Ψ).

Variability (Z) is the proportional constant as shown below.

$$(7) \qquad Z = \Phi/\Psi$$

Logarithm of both sides are taken and differentiated,

$$(19) \qquad (d\Phi/dt)/\Phi - (d\Psi/dt)/\Psi = (dZ/dt)/Z$$

The variability induces that the remainder of insecurity variation rate and inconstancy variation rate is variability variation rate. Equations (2) and (3) are substituted into equation above then equation below is obtained.

$$(20) \qquad (dX/dt)X/\sigma^2 - (dV/dt)V/\tau^2 = (dZ/dt)/Z$$

X is deviation from point (a) on axis x and $V = dX/dt$. Z is a constant. Then equation below is obtained.

$$(21) \qquad d^2X/dt^2/\tau^2 - X/\sigma^2 = 0$$

Let $\tau/\sigma = \omega$ then the dimension of ω is T^{-1}, which is the dimension of angular velocity.

$$(22) \qquad d^2X/dt^2 = \omega^2 X$$

The objective point at $x = a$ is expected to be revolving and has positive acceleration which causes revolving centrifugal force.

(2) Continuity and revolution

Continuity is the stabilized existence. The continuity (H) is the product of existence probability (Φ) and persistence probability (Ψ). Since they are anti-proportional the continuity (H) is the proportional constant. When an actual point (A) stays still at position (a) on axis x the continuity is,

(16) $$H = \Phi * \Psi$$

Logarithm is taken on the both sides of above equation is differentiated.

(23) $$(d\Phi/dt)/\Phi + (d\Psi/dt)/\Psi = dH/dt/H$$

Equations (2) and (3) are substituted to equation above and then equation below is obtained.

(24) $$(dX/dt)X/\sigma^2 + (dV/dt)V/\tau^2 = dH/dt/H$$

X is deviation from point (a) on axis x. and $V = dX/dt$. H is a constant. Hence equation above becomes equation below.

(25) $$d^2X/dt^2/\tau^2 + X/\sigma^2 = 0$$

The ratio of τ/σ is revolving angular velocity (ω). Equation above is an oscillating function. The mass point is revolving at the point a. Then equation below is obtained.

(26) $$d^2X/dt^2 = -\omega^2 X$$

Continuity is revolving. The objective point at (a) is expected to be revolving and has negative acceleration which causes revolving centripetal force. Centrifugal force by equations (22) and centripetal force by equation (26) keep the objective point stable. The stabilized objective point still has the angle velocity, which means it is revolving on point (a). As equation (22) indicates the revolving objective point at point (a) has positive acceleration which causes revolving centrifugal force. And as equation (26) indicates the objective point has negative acceleration which causes revolving centripetal force. The both forces stabilize the point. Rotating events actualize a stabilized substantial point.

(3) Substantiality and energy

The actuality of physical events is usually perceived with scholar values. But the values should be considered as the most probable statistic values which are determined by the balance of variability and continuity. The objective point at $x = a$ is expected to be essentially revolving and stabilized.

But the insecurity of existence is very large the existence probability distribution becomes close to flat and the inconstancy is also large at the same substantiality. Then the revolving becomes waving. And due to its variability it has positive direction. Continuity is also waving but toward the other direction. Hence this objective point is stabilized. These waves are energy waves. The wave strength of both directions is the same since the wave frequencies of both directions are determined with (τ/σ) which are the same in both directions. Substantiality (Θ) is the characteristic of an event and is its essentiality or the value of existence. Substantiality is a constant which is independent to time and determines the energy strength. Energy depends not only substantiality constant but also insecurity and inconstancy which will be discussed later. The intensity of energy is brought about by wave height which depends on the group size of wave sauce.

2. Stabilization of space

A space exists conceptually based on one fundamental conceptual point and three fundamental directions. This conceptual space is the most fundamental frame for the existence of objective points and its movements. Hence this conceptual space can be the framework of objective existences. An objective point can be placed at one conceptual point in the conceptual space. This objective point can have three fundamental directions which form an objective space. This objective space has instability depending on the fundamental objective point. That conceptual space is the reference frame for this objective space which can be also the reference frame for another objective space. The objective points in these spaces have no mass. But they are actual existents and have positions with various deviations and motion with various fluctuations referring to corresponding points in the reference frame. Time in a substantial space makes existence change. Points in the space have some instability, therefore, they have existential probability and persistence probability, which induce continuity and variability. Hence it has substantiality. The substantial space consists of this kind of objective points together with subjective points, therefore, it is substantial and has instability. Variability of substantial points follows equation (22) and makes centrifugal acceleration. Continuity follows equation (26) and makes centripetal acceleration. Both equations make the stable existence of space. The points in it are stable and are revolving. The space which is consisted of this kind of substantial points is named substantial space. Substantial spaces are stable.

3. Stabilization of moving points

If there would be an objective point at a position (b) in axis x whose existence reliability would be lower than that of position (a), then the point makes motion toward position (a) on axis x.

Position (b) is regarded as deviation (X_0) of position (a). Centripetal force from position (a) affects to the objective point at (b).

When the objective point is moving along x-axis the equation (10) above can be changed as follows.

(12) $$[d/dt+(\tau/\sigma)][d/dt-(\tau/\sigma)]X = 0$$

(13) $$dX/dt-(\tau/\sigma)X = 0$$

Since equation (13) diverges to infinite. It is a destructive process and is not realistic. This is a stabilizing process and here a converging equation should be applied. The solution of equation (12) shall be as follows.

(14) $dX/dt+(\tau/\sigma)X = 0$

The solution of this equation is following abducting function where X_0 is initial value of deviation X at position (*b*).

(15) $X = X_0 * \exp[-(\tau/\sigma)t]+C$

C is a constant

The deviation (X) reduces with the attenuation coefficient (τ/σ) and converges to 0, which means that the amplitude of wave function (26) converges to 0 and the existence reliability of point (*b*) disappears and the existence reliability of point (*a*) increases. But this is not realistic solution since continuity must work on it. The objective point makes motion according to the variability which is equation (14) and stabilizes according to the continuity which is equation (25). According to the substantiality the sum of these two equations shows the motion.

Equation (14) shows the motion from position (*b*) to position (*a*) due to the variability. Equation (14) can be changed as follows.

(27) $2\omega dX/dt+2\omega^2X = 0$

The continuity contributes to the motion. Equation (26) is changed as below,

(28) $d^2X/dt^2+\omega^2X = 0$

According to equation (18) the substantiality of motion is the sum of equations (27) and (28), which is the equation below.

(29) $d^2X/dt^2+2\omega dX/dt+3\omega^2X = 0$

The solution of the equation above is as follows.

$$(30) \qquad X = \exp(-\omega t)*[X_1*\cos(2^{1/2}\omega t)+iX_2*\sin(2^{1/2}\omega t)]$$

X has to be a real number, therefore, $X_2 = 0$.
When $t = 0$, X is initial deviation X_0, therefore, $X_1 = X_0$
Then equation (31) is obtained which is an abducting wave function.

$$(31) \qquad X = X_0\exp(-\omega t)\cos(2^{1/2}\omega t)$$

Wave amplitude X reduces and becomes close to 0 and the objective point is stabilized. The stabilization of objective points follows this process exactly as the gravity law of mass points. On the movement of objective points equation (31) is an approximation since (τ/σ) is regarded as constant. Actually the existence should increase and the persistence should decrease, which means (σ) decreases and (τ) increases. The point-fusing makes existence and persistence change. This process shown above may roughly explain the process of objective point-fusing. When this kind of fused objective points is a huge lump the equations (22) and (26) are also applicable to this "material" which yields material wave by De Broglie. Energy is very high frequency spatial or material wave.

H. Fused objective points

1. Objective point fusing

An objective point has position and motion. Hence it has insecurity (σ) and inconstancy (τ) like a mass point. Actual stable stay of an objective point at one position means the objective point has the statistically most probable presence which implies a little improbability. An objective point at a position bears insecurity (σ) which hardly changes in point-fusing and its motion bears inconstancy (τ) which increases in fusing and decrease persistence.

A restrict space has infinite number of Euclidean points. At the same time it has very large number of objective points. Both of them have no dimension. These objective points have all sort of substantiality. But their existence has uniformity since they are just points and have no dimension. They have non-uniform persistence. This condition is brought forth by the fusions and dissolves of objective points. The following processes can be considered. The multiple point fusing is expected to increase the variability ($Z = \Phi/\Psi$) which is the ratio of existence (Φ) and persistence (Ψ). Existence change little in point fusing. Persistence decreases, which means inconstancy (τ) increases. Objective points make fusing and multiple fusing decreases persistence (Ψ), which induces the higher variability and the higher substantiality. And in consequence it releases the higher frequency energy wave as equations (22) and (26) show. The ratio of τ/σ is revolving angular velocity (ω). $\tau/\sigma = \omega$ then the dimension of ω is T^{-1}, which is the dimension of angular velocity. Equation (22) is oscillating function from variability which releases energy wave toward the positive direction. And equation (26) is oscillating function from continuity which releases energy wave toward negative direction. Hence the fused objective point is stable and keeps its position.

The incidence of multiple objective point fusion is low. The majority of those points stay singles or low times fused points. Hence point-fusions occur according Poisson provability distribution which means highly repeating of point-fusion occurs not often. It can happen that fused objective points may dissolve, which induces lower energy.

2. Assemblage

Various fusing of objective points can occur in space. Various large number of objective point fusing form the assemblage of fused objective points which is still a single point and has its own substantiality. The occurrence follows Poisson probability distribution. Therefore, the occurrence of low level assemblage is much higher than that of high level ones. Very low level assemblages generate weak energy waves. They are not sensible and are designated here intangible energy. Lowest tangible energy is heat. Energy which is stronger than heat is designated here tangible energy.

3. Cubage

An assemblage consists of objective points and the points have no dimension, therefore, an assemblage has no dimension and is still a point. But multiple active assemblages require some space. The existence of very small size congruous subspaces can be considered where the composition of various assemblages is similar. They are designated here cubage which is the minimum subspace for considering energy generation. In a cubage there occur multiple levels of objective point assemblages which have substantiality. The formation of assemblage levels follows Poisson probability distribution. Majority is intangible energy level. Hence the occurrence of heat energy level is the highest among tangible energy level assemblages in a cubage. The occurrence of assemblages of lower than the minimum tangible energy level is neglected here. Hence the minimum tangible energy which is heat level represents the cubage contents. In a cubage the peak existence probability of tangible energy is the heat energy at zero Kelvin degree, which represents cubage substantiality. The tangible energy above heat level is not much. But higher energy level substantiality is also available in a cubage.

4. Intangible energy

Contingent unifying of objective points with existence probability distributions can occur. They are assemblages. They have casual differences of existence and persistence probabilities which induce all sorts of substantiality. The substantiality of assemblage makes revolution and releases waves which are energy. The energy wave frequency which is below zero Kelvin degree is intangible energy. Inconstancy increases due to increased substantiality in the statistical process of assemblage fusing, which induces higher energy wave. In a cubage fused various assemblages release various higher frequency quality energy waves. Majority of assemblages in a cubage are the intangible level assemblages. Energy has quality and quantity. Quality depends on wave frequency and quantity depends on wave height or the amount of wave sources. The energy frequency below zero Kelvin degree is hardly perceptible but can modify various physical events. Examples are energy waves which can easily modify the existence of air or water with air or water waves. High energy frequency is generated from high inconstancy of substantiality from the multiple repeating of assemblage fusing. Incidence of assemblage fusing depends on Poisson distribution. Hence majority of energy waves in space is non-perceptible under heat energy level. Low level occurrence of fusing is expected much more than high level on Poisson distribution. Strength of waves or wave frequency depends on increased substantiality which is expected on the increased fusing of assemblages or on the cubage merging.

5. Multiple cubage merging

It is considerable that a certain volume of space carries a certain number of cubage which contains assemblages of almost similar compositions. In a cubage there is enormous number of assemblages and the existence probability among tangible energy levels is the highest at heat energy level with sharp decreasing curve. Each cubage can be considered having the almost same composition of assemblages, which means energy level is represented with the minimum tangible energy of heat. Cubage is considered having almost the same composition of assemblages in average.

Cubage merges often and forms various assemblages according Poisson distribution. In a merged cubage assemblages fuse each other and have new assemblages of higher substantiality depending on the number of fusing times. The higher fusing times afford the higher substantiality which yields the higher energy frequency. Multiple merged larger assemblages generate stronger energy. The strength of energy means the wave frequency which depends on the substantiality and the quantity of energy depends on the amount of the fused assemblages of the same level.

I. Tangible energy

1. Cubage merging and high energy

Cubage merging which makes stronger substantiality shall be considered. The high number of cubage merging which has low probability in Poisson probability bring forth highly repeated assemblage fusing and releases high frequency tangible energy. Those fused assemblages have contingent differences of existence and persistence probabilities which results in various tangible energy waves.

The standard assemblage is considered here which keeps fundamental substantiality (Θ_1) and releases the energy waves of zero Kelvin degree.

And then Poisson distribution curve of various expected sizes substantiality must be evaluated which are due to the assemblage fusing times (λ) of standard substantiality ($\lambda * \Theta_1$). The postulated fusion times (k) evaluates each expected fusion times (λ) of the standard substantiality in the space. P stands for the probability of postulated fusion times (k) at each expected fusion times (λ) which determines P(k, λ) and will be evaluated with the equation below.

(32) $$P(k, \lambda) = \lambda^{k} * \exp(-\lambda)/k!$$

λ is expected number of fusion times
P is provability of the fusing times
k is postulated fusion times at each λ

Objective points have substantiality which makes wave. Wave has strength and intensity. Strength depends on wave frequency, therefore, the modal value of λ determines energy strength. And the intensity depends on wave height which is determined by the amount of similar substantiality.

2. Generation of energy

Multiple merging of cubage causes higher substantiality which is induced by the large number of fundamental assemblage fusing. High multiplication (λ) induces high substantiality (Θ) and in consequence generates stronger energy depending on Poisson probability.

Usually at the beginning the existence provability ($P(\lambda)$) increases as fusing times (λ) increases. Then $P(\lambda)$ makes the peak and thereafter ($P(\lambda)$) decreases. But when the modal value of (λ) is very small the peak comes close to the beginning, therefore, such small modal values of (λ) is considered not existing actually. As the value of (λ) increases the existence probability curve of $P(\lambda)$ distribution becomes close to the Gaussian probability distribution. Hence (λ) value has a peak. The peak or the modal value of λ determines the existence.

The mean or modal value in Poisson or Gaussian distribution reside at $k = \lambda$. Then the modal existence probability of fused substantiality is at $k = \lambda$ and the modal existence probability ($P(\lambda)$) of substantiality is,

$$(33) \qquad P(k, \lambda) = \lambda^{\lambda}/\lambda!*\exp(-\lambda)$$

Objective points have substantiality which makes wave. Wave has strength and intensity. Strength depends on wave frequency, therefore, the modal value of λ determines energy strength. And the intensity depends on wave height which is determined by the amount of points with similar substantiality in a cubage. Something happens which makes λ value too large or too small the existence probability of substantiality disappears, which is accompanied by the extinction of energy.

3. Tangible energy modes

Energy strength depends on the substantiality. Substantiality consists of existence and persistence and therefore it is the product of insecurity and inconstancy. Energy modes at the same level of substantiality can be different depending on the values of insecurity and inconstancy. Hence tangible energy can be classified as below.

A. High insecurity point releases wave energy as follows.
  1. With low inconstancy releases low frequency waves as temperature
  2. With high inconstancy releases high frequency waves as light

B. Low insecurity points become quantum or particulate with or without mass
  3. With low inconstancy releases low rotation grain energy as quanta or quarts
  4. With high inconstancy releases high rotation particle energy as meson or mass

4. Heat

Minimum tangible energy is heat energy. Substantiality of assemblages makes revolving and releases energy wave toward +/− two directions. These assemblages face various directions and because of high insecurity they scatter around, therefore, heat energy is filled in a space.

5. Light

Light is wave. Light has very high frequency wave. Red light has about 450 billion cycle per second. And blue light has 550 billion cycles per second. Usual visible light has the frequency of 430 billion to 770 billion cycles. They are released from the substantiality of high insecurity and high inconstancy. Light also has characteristic of particles which are quanta.

6. Quantum

Multiple sorts of substantiality points in cubage release non-uniform high frequencies of lights. Various light high frequencies overlap and makes irregularity of wave frequency. Low frequency portions and very high frequency portions occur. The very high frequency portion solidifies and becomes a quantum.

7. Quarks meson

Low insecurity points with low inconstancy substantial point releases particle energy as quarks or mesons.

8. Mass

Low insecurity points with high inconstancy release particle energy with mass.

J. Substantiality, existence and energy

1. Force and impetus

Substantiality has existence which has some instability. The existence has the probability or capability to return to normal topological condition, which is the force as equation (34) indicates. The positional integration of force is the existence. This positional integration of force is energy.

$$(34) \qquad dE = F^{*}dx$$

Hence the change in energy is defined as force multiplied by the change in existence.

The independent variable of persistence probability is velocity. This variable should have some fluctuation which is temporal instability and makes acceleration. Persistence is capable to have acceleration to recover normal persistence, which is force of substantiality. The temporal integration of force is the persistence as equation (35) indicates.

$$(35) \qquad dP = F^{*}dt$$

The change of impetus also generates force. Force multiplied by the change in time is change in impetus.

High variability has tendency to decline itself and makes stronger force and temporal integration of force is impetus. Impetus makes the objective point move. It may move as heat which is energy. It may fly straight as light which is energy. Then it makes the variability lower. Temperature which is a shape of energy represents the variability of substantiality. As temperature comes down substantiality becomes more stabilized, which means substantiality increases continuity and decreases variability and its impetus becomes lower. From equation (35) F is equivalent to dP/dt, so plugging it in for F of equation (34).

(36) $$dE = dP/dt * dx$$

Change in position over change in time becomes velocity.

The energy instability which is change in existence becomes change in impetus multiplied by velocity.

(37) $$dE = dP * v$$

2. Substantiality and impetus

Substantiality has existence and persistence. A stable state of space is composed of large number of objective points of the same substantiality, which are well balanced and have no movements. The stability depends on the flat distribution of substantiality in space.

An objective point with different substantiality can happen to be in space and possible positional uneven value of substantiality can occur, which has the tendency to induce the movement for the stabilization. But this substantiality has all the same force in all directions around this objective point, therefore, this objective point is still stable.

When there is nearby another objective point with different substantiality the impetus from both points influence one another and move toward the other point, where variability of both substantiality of both objective points decrease. Those objective points may fuse and gain new substantiality. In this way the objective points are reacting with surrounding points in space and gain the positional substantiality balance.

Spatial uneven distribution of substantiality makes objective points flow where its variability changes. The flow of substantiality is impetus. The uneven substantiality distribution makes flow and change in impetus, which induces change in existence. Substantiality is not constant. The effect of substantiality change is given in all directions. Change in substantiality induces change in impetus. The flow (v) of substantiality effects change in existence and impetus as equation (37) indicates.

Positional energy which is the existence level difference (dE) is the temporal change in impetus which is force times the change in position. Motion energy is the change in impetus time velocity.

3. Solidification of substantiality

Substantiality of objective points has variability and continuity. Variability induces revolving centrifugal strength and continuity induces revolving centripetal strength. Objective points are revolving, not stable points. The revolving requires a tiny space. The objective points with extremely high variability and extremely low continuity of substantiality have extremely high frequency revolving centripetal and centrifugal strength. A group of those objective points connect one another and is solidified, which is mass or material.

Those solidified objective points restrain each other and gain their stabilized variability and continuity. Such solidified objective point is a mass and obtains its own substantiality of high stability which has high continuity and low variability.

4. Mass

Mass is the solidified shape of substantiality, which means it has the high continuity but it should have a little variability. Hence mass has solid existence and is always integrated with various movements according its condition, although it has extreme high continuity and its variability is extremely small. A mass is a different existence from the just gathering of objective points. It gains its own variability, continuity and substantiality. Its substantiality has very low variability and high continuity. The material of this existence is a group of high frequency substantiality objective points whose revolving frequency is extremely high and is not progressive. The material wave length of an electron is less than one thousandth of that of visible light. They are just a compound of revolving points.

5. Modified energy

A mode of energy sometimes modifies the actuality of other mode of energy. An example is hot water. Heat energy modifies the substantiality of a solidified energy of water. Large intensity condition of heat energy overheats the heat condition of water and generates the humid shape of water substantiality. The merging of both shapes of substantiality takes time.

The motion of substantiality changes impetus and energy. This change of energy can disappear or changes the condition of substantiality in time. Substantiality has the tendency to increase continuity and decrease variability. But an unstable condition of energy may increase variability and decrease continuity. The substantiality may change its shape.

6. Mass and its modified activity

A mass is the solidified shape of objective points and it has its own much stabilized substantiality which can be modified with various elements of energy. The quantity of object points or combined substantiality size is expressed with the quantity of mass (m) which is a constant substantiality and its flow (v) make the impetus which is its energy flows (dE/dt). The movement of mass forms momentum which is impetus and its integration becomes energy.

Simple constant movement of a mass is momentum.

$$P = m * v$$

Its temporal integration is energy.

$$E = m*v^2/2$$

Force is the product of mass (m) by acceleration (dv/dt).

$$F = m*dv/dt$$

7. Wave energy

Tsunami is a large amount of the substantiality of water modified with very high intensity of low frequency low strength intangible energy of wave. Its substantiality has low variability and very high continuity of water wave. Its movement has strong impetus which can make water wave much stronger. Its impetus increase is not due to water substantiality but to wave energy intensity. The actuality of water is high continuity and low variability substantiality. The flow of this substantiality increases its impetus with the impetus from enormous amount of intangible wave energy. Its quantity depends on the quantity of water substantiality which is the amount of water.

8. The law of energy conservation

Equation (37) indicates that an event with no movement nor change in impetus in total has no change in existence which indicates the event has had no change of existence value in total. Existence value is proportional to energy. Hence there is no change of energy which mean energy is conserved.

$$(37) \qquad dE = dP*v$$

Substantiality consists of variability and continuity and generates impetus.

Each object point has its own impetus which keeps the stability. When some event happens in the space the objective points in space are reacting with surrounding points in space and they gain or lose substantiality. Substantiality value of each objective point may be different before and after the event. When the event is stabilized the total impetus has no change unless the outflow of substantiality happens.

$$\text{(35)} \qquad dP = F^{*}dt$$

Equation (35) shows events without force has no change in impetus. Equation (37) shows when there is neither motion nor the change in impetus there is no change in energy. The law of energy conservation can be applied to this situation.

No change of total impetus makes no change of total existence as equation (37) indicates. Existence value is proportional to energy, therefore, there is no change of energy which mean energy is conserved.

K. Hypothetic existence and electric energy

1. Hypothetic state of energy

The standard deviation in equation (2) defines insecurity. As the standard deviation increases, the existence reliability of an objective point decreases and the substantiality of the point decreases and then disappears with infinite value of insecurity. Equation (38) shows that the centripetal force disappears with infinite insecurity. This state is hypothetic. A stable flat distribution of existence reliability with positive or negative infinite insecurity is conceivable. The existence probability of energy becomes flat. Hypothetic existence with a negative insecurity is also conceivable.

$$(2) \qquad \Phi = \exp[-X^2/(2\sigma^2)]/[\sigma(2\pi)^{1/2}]$$

The density of existence probability of energy E is presented by the distribution of equation (2). With a negative insecurity central force occurs with equation (38) and the hypothetic energy excess (E). From equation (2) the central force (F) is proportional to $[1/\sigma(2\pi)^{1/2}]$.

$$(38) \qquad F = E/[\sigma(2\pi)^{1/2}]$$

And hypothetic energy excess (E) and energy deficits (−E) with negative insecurities can be considered in the hypothetic state. The centrifugal force occurs to an energy excess and the centripetal force to an energy deficit with hypothetic negative insecurity.

2. Hypostatic forces

Equation (2) is applied to the hypothetic point with negative reliability.

X shall be deviation from the expected point of a hypothetic existence which is energy excess or energy deficit.

The centrifugal force works to an energy excess. And the smaller absolute value of the standard deviation is the stronger is the centrifugal force. The energy deficit draws energy and the deficits have tendency to restore. The excess has tendency to decay. When equation (38) is applied to the excess where F is negative and is centrifugal force. Frailty of the excess at the expected point ($X = 0$) is $1/[(2\pi)^{1/2}\sigma]$. The higher this absolute value is the frailer is the excess, therefore, the excess has the more pressure to be flattened. Frailty increases as the absolute standard deviation (σ) decreases, therefore, the reciprocal ($1/\sigma$) expresses the tension to decay. The centrifugal force is proportional to the tension ($1/\sigma$). From equation (2) the central force (F) is proportional to$[1/\sigma(2\pi)^{1/2}]$.

Proportional constant E is considered as the excess energy. When E is a unit of energy the probability distribution is the same to the excess energy incidence at deviation X. $-F$ is the tension in the center of distribution.

3. Substantial existence and hypostatic state

A particle is a substantial existence with energy. An example is a neutron which is a substantial existence. A neutron can become a proton and an electron which are consisted from energy. When a neutron releases an electron the new existence which is proton can't always gain perfectly balanced energy. An energy excess which is a hypothetic existence can be left in a new proton. At the same time the electron has an energy deficit which is also hypothetic existence. These proton and electron generate Coulomb force and the electron exists rotating around the proton with a certain angle velocity. High intensity energy which is energy of high frequency wave quantizes or particulates. Particles have their own proper energy levels and energy excess or energy deficit can happen. These excess or deficit is hypostatic state of energy.

4. Coulomb force

Energy mound is fragile and causes centrifugal force in its vicinity. A mound of energy E which is considered as +Coulomb makes the centrifugal force. The energy incidence at the center of energy E is $E/[\sigma(2\pi)^{1/2}]$ and its centrifugal force is obtained applying equation (38). When there is an energy hollow (−K) at the deviation X of the existence probability distribution of energy E. Then the existential incidence of energy E is modified due to the existence probability distribution of energy (−K). The centrifugal force of energy E at its center is given by the product of energy E and the energy incidences of energy (−K) at the center of energy E. In this case σ is communal and F becomes negative. Equation of the Coulomb force is obtained as follows.

$$(39) \qquad F = E\exp[-0^2/(2\sigma^2)]/[\sigma(2\pi)^{1/2}]$$

$$*(-K)\exp[-(-X)^2/(2\sigma^2)]/[\sigma(2\pi)^{1/2}]$$

$$= (-K)E\exp[-X^2/(2\sigma^2)]/(2\pi\sigma^2)$$

When X is large enough the approximation is done applying Tailor's expansion.

$$= (-K)E[2\sigma^2/(-X)^2]/(2\pi\sigma^2)$$

$$= (-K)E/(\pi X^2)$$

The energy incidence (F) at X = 0 becomes negative which is centripetal force.

$$-F = KE/(\pi X^2)$$

This equation can be changed as follows

$$-F = E/(\pi^{1/2}X)*(-K)/(\pi^{1/2}X)$$

Force from Coulomb has been defined as follows

$$-F = Q(-Q)/X^2$$

$$-F = Q/X*(-Q)/X$$

Hence Coulomb (Q) of energy (E) is,

$$E = Q\pi^{1/2} \tag{40}$$

Coulomb has the same dimension to energy.

Epilogue

All existents have continuity. But none of those existents continue forever. All of them have more or less variability, which is substantiality. All existents have substantiality. On stabilization principle the substantiality is energy. All events of existence are consisted of energy. There is energy with high variability or energy with high continuity. All existing events depend on energy and their change is due to change of energy. The existence of an event is the existence of energy. The change of an event is the change of its own energy which can be the effect from energy from outside.

安定化原理に基づくエネルギー発生

プロローグ

質点の存在はその存在環境との調和に基づいて存在すると考えたのが安定性理論である。存在環境が変化するとその新しい環境に質点が変化し調和する理論について安定性理論としてまとめた。しかし、質点には安定な存在だけでなく変動をもたらす運動がある。この運動こそ安定をもたらす安定化であり、その安定化とは存在確率の上昇である。これこそ認め易い安定化原理であるとして考え直すとエネルギー波の発生まで説明できる。存在とはエネルギーの安定化であることを追求した。

目 次

A．空間と存在性

1．概念的存在と実在的存在

一般に考えられる存在概念は思考上安定なスカラー量で考えられる。その存在は主観的である。しかし自然の物理的存在は実在的である。その本質は客観的であり人の考えとは独立している。人は自然的存在物をほとんど安定であると思う。空間は思考上安定な三次元であり、次元も不安定性もない無数の概念的な点から成ると考える。空間は思考上三次元で安定な概念でありこれら無数の概念的な点を有するのが本質である。

しかし空間には時間があり、それが安定性を妨害し、それら安定な存在に客観的な不安定性をもたらす。物理的現状を有する点は本質的には偶然がもたらす自然法則に従う。現実の客観的な点は偶然性がもたらす実在的な産物であり、偶然性に基づく法則に制御されている。偶然性には安定と不安定な存在性がある。偶然性は安定な存在の周囲とのバランスを壊し不安定な存在にする。このバランスこそ自然の不安定要素である。

2．主観点及び客観点

点の概念はユークリッドにより定義されている。それには長さや空間は伴わない。体積も実質性もない。それはただ概念的に正確な位置を示す。それらの間の主観的な関係を示している。それらの点は主観的で物理的な事象への関与はしていない。

しかし幾つかの点は物理的事象に関与している。それらには客観的に不安定な存在と運動があり実在的な存在である。例として質点がある。

それには実在的な存在と運動のある不安定性がある。これら物理現象に関与する点は実質性があり客観点と名付ける。

3. 主観的及び客観的空間

空間は三次元で無限数の点から成るとユークリッドは定義した。この近辺の空間は三次元であり本体論的に定義された無限数の点から成るとの概念で主観的である。宇宙の大空間は存在するがそれは十分には分かっていない。それは普通三次元と考えられている。その性格はまだ大いに追求されるべきである。いろいろ多次元空間は各種研究に利用されている。それらの本質は概念論である。安定な三次元の空間も主観的概念論である。

この近辺の空間は時間的不安定をもたらす三次元空間と認識されるべきである。実在的空間には位置と運動があり、一次空間に対して客観的不安定性がある。その一次空間は主観的空間でも客観的空間でもあり得る。だから客観的空間内のすべての点の存在性と持続性に不安定性があり、客観的空間全体に実質性をもたらす。客観的空間は実質性のある存在である。客観性の空間は大多数の実質性のある点と無限の時間から成る。この近辺に現存する客観的空間は温度空間であり、それは宇宙空間内の客観的な点で構成されていることによる。

4. 二次空間

空間にはその中に二次空間が存在し得る。二次空間の参照枠は一次空間である。参照枠となり得る一次空間は実在空間でも概念空間でもあり得る。だから客観的二次空間には少々位置と運動の不確定性があり、それゆえ実質性がある。実在空間である二次空間は三次元で固有の時間がある。時間が不安定にする。二次空間座標系の一点に存在する客観点には不安定性がある。客観点が一点に留まるのはその点が最も適切な位置であるからである。しかしその点には不適正性もあるゆえ、その点には存在性と持続性に少々不安定性がある。

B．実 質 性

1．時　間

存在の実質性を最も重要な要素である時間の要素抜きでしばしば考えられることがある。客体の存在は時間を用いて三次元内で考えられるべきで時間の概念が必要である。空間内で検討される時間は経過時間である。ここで時間と呼ばれる経過時間とは一空間内で一瞬間から次の瞬間までの経過時間である。実在空間にはそれぞれそれ統計量である固有の時間がある。客観的空間では動的事象において相対性が生じる。時間は各空間においての独立変数である。例えば客観点の基本軸方向への移動距離が時間の従属変数となる。空間と時間が存在の枠である。物理的事象を表すには空間と時間が必要である。客観的点の存在を表すのが空間時間枠で、それはその客観的点が属する実在空間である。

2．概念的及び実在的点

事象の存在はユークリッド空間と呼ばれる概念的な時間なしの安定な三次元空間内で考慮される。点は位置を示し次元を持たない。それはユークリッドの点である。その存在は概念的でありここでは概念点または主観的点と呼ぶ。ユークリッドの点は座標系の一点に存在し安定である。

しかし実質性即ち客観性の点は少し不安定性がある。それには散在性や揺動性がありここでは実在点または客観点と呼ぶ。例としてはボールの重心点がある。それは安定でなく散在性や揺動性がある。それが散在性や揺動性のある実質性の点で客観点である。質点が実在点の例である。質量のない質点が実在点であり客観点とも呼ぶ。実在する客観点はいろいろな実在価を有する。物理的事象を示す客観点には存在性と持続性があり位置と運動を示す。安定性を示す時間軸は各空間に存在する。事象の存在はユークリッド空間と呼ばれる時間のない三次元空間で検討されてきた。事象の存在は時間軸の存在する客観的空間にて検討されるべきである。客観点には位置と運動があり、それらは時間が関与する速度と加速度から成る。

3．実在空間

客観点の位置と運動を表すには尺度または座標系が用いられる。それはその客観点の属する空間は安定な状態として存在するとして用いられている。それゆえその空間自身には位置や運動はない。概念空間は最も基本となる点、原点と基本の３方向から成る。最も基本となる空間は概念的であり無限数のユークリッド点から成る。最も基本の客観的な点は概念空間の一点と相関して客観的な二次空間を形成する。概念的即ち主観的空間を基に散在性や揺動性を有する点の集合で客観的即ち実在的二次空間を形成する。

多数の客観点から成る実在的即ち客観的な空間は存在と持続の確率分布を有し、その原点は客観点であるから主観的空間内で不安定である。実在的空間は客観点を含めていろいろな実質性の存在事象を保有する。偶然性がそれらの存在に散在性や揺動性を起こさせるゆえ、それらの存在は変動性や存続性を受け結果として実質性を持つ。

客観的空間には固有の時間と座標がある。しかし、主観的空間群では共通の時間であり不安定性のない存在であるゆえ実質性はない。そのうちの一つが最も基本の空間となる。一基準点と三基本方向の空間は動き得る不安定な二次空間を与える。空間には可動性で不安定な二次空間がある。二次空間はその中に更なる二次空間を持ち得るが最小の二次空間はサイズ０空間ではない。サイズ０の空間は点である。二次空間内のすべての客観点は散在性による存在確率分布や揺動性による持続確率分布を有する。

非常に小さな空間即ち客観点の周囲のミクロ空間（微泡）にも実質性はある。この客観点はその微泡を表すことであるが実在点と呼ぶ。この実在点の微泡の大きさは決まっていない。

C．安定化原理

客観点は統計的確率に基づいて安定化する、それが安定化原理である。安定化とは統計的確率が上昇することである。統計的確率分布の局所的最大の状態が安定をもたらす。位置と運動の確率状態が局在的に最大となっている状態が安定である。位置と運動はスカラー量ではなく統計量として扱わなくてはならない。単純な例として1mの測定値である。この1mは概念的な1mで、この1mにそうでない確率も共存していることを意味する。即ち存在性がある。事象はより有り得る方に変化する。事象の存在と持続はそれらの確率確度に従う。よってそれらの値は統計的に表示されるべきである。事象はその確率に基づいて変化する。統計的な確度が上昇すればその存在性は上がる。持続性も必要である。持続確度が上がれば持続性は上昇する。存在性と持続性で実質性を成す。実質性の本質は陰性エントロピーと言える。

存在性変化の一例を水の温度変化で示す。0度の冷水を25度の空間に放置するとやがて冷水は25度になる。これは安定化である。

スカラー量を用いた概念的安定化の解決法を示す。初期温度は0度で最終温度が25度である。概念的解決法では水の温度上昇は温度差に比例するとして下記に示す式となる。

xは水の温度、tは時間、aは水の最適温度25度、bは変化率である。

$$bx = \mathrm{d}(a-x)/\mathrm{d}t$$

この等式は下記の如く変換できる。

$$\mathrm{d}x/x = -b*\mathrm{d}t$$

この等式の解は下記の如くで安定化の減衰式である。

$$x = a^*[1-\exp^*(-bt)]$$

これらの式は概念的である。この x はスカラー量である。これらの式は水の温度の安定化を表している。だから安定性原理から後のＦ章（変動性と存続性）で述べる。その定理である変動性を用いて導くことができる。その場合この x は統計量であり存在性と持続性の確率分布を有する。例の存在性は次のように表せる。次の式が存在状況を示している。

$$(1) \qquad \Phi = \exp[-(x-a)^2/(2\sigma^2)]/[\sigma(2\pi)^{1/2}]$$

σ は Φ の標準偏差　$\sigma > 0$

x はグラス内の水温で気温により変化する、そして a はこの存在の最適温度。

簡便のため X は最終最適値からの偏位を用いる、V は持続性で dx/dt。それらは統計量である。X は最適点（a）からの x 軸上の距離、それは $X = x-a$。

$$(2) \qquad \Phi = \exp[-X^2/(2\sigma^2)]/[\sigma(2\pi)^{1/2}]$$

$$(3) \qquad \Psi = \exp[-V^2/(2\tau^2)]/[\tau(2\pi)^{1/2}]$$

これらの式を変動性の式 $\Phi^*\Psi$ に代入して得られる結果は $dx/dt = -bx$。

この等式は上記の概念的解決法で用いられた式に同じである。この経過は後で変動性の項で説明される。

存在と持続は統計的に定まるものでありスカラー値では定まらない。

いかなる事象でも客観点には確率分布を伴う存在性と持続性があり、その確率に基づいて変化する。客観点はこの信頼度を上昇させ安定になる。高信頼度の状態とは締め付けのない状態である。実際の存在には幾ばくかの不確実性を伴っている。客観点は客観的存在であり、その存在は常に時間に影響されている空間内の一点であり、その存在性は常に持続性を伴っている。客観的存在とはその存在性に持続性を伴う存在である。空間が存在性を与え時間が持続性を与える。上記の例の事象での存在性は散在性を含む温度 x であり、持続性は揺動性を含むその時間的変化 $\mathrm{d}x/\mathrm{d}t$ の V である。

安定性原理は次の２構成要素からなる。それらは公理である。

1．散在性；事象は存在する。存在には統計的な散在性がある。
2．揺動性；事象は持続する。持続には統計的な揺動性がある。

D．客観点の存在と持続

1．存在の散在性と揺動性

⑴ 存在の不確定性

存在である質点には位置と運動がある。質量のない客観点も同じである。客観点には位置と運動がある。その位置には少々の不確実性があり、それが散在性である。その運動には少々のぐらつきがあり、それが揺動性である。

人の考える点は概念点である。それはどこでも正確に概念空間内にて位置決めできるし自由に運動できる。その運動は完全に安定している。

自然に基づいている客観点は最も単純な存在である。それは人の考えるような正確な位置決めや動きはしない。それは人の考えには関与せず自然の法則に従う。自然の法則に従う客観的存在は偶然性が関与する。物体上の一点はある程度不安定で人の考えとは別である。

物理空間では概念点の存在と偏位や変動のある客観点の存在には本質的相違がある。客観点には位置と運動がある。その位置にはずれがあり得る。それを偏位と呼ぶ。そして運動には揺れがある。それを揺動と呼ぶ。

⑵ 事象の安定化

事象は存在し安定化する。客観点の存在は一つの存在である。一次量と二次量がある。それらの量には双方の量に統計的な不安定を示す変数が少なくとも一つある。例えば電気量に関しては電気量が一次量でそれは電圧で決められるが変化もする。電気量の時間変化は電流でありそれが二次量であり電圧で決められる。共通変数は電圧である。二次量は一次量の変動である。他にも例として質点がある。質点は一点上に留まっている。これは一次量である。この存在はあるべき点から移動する可能性がある。それが運動である。この運動が二次量であり距離を含む量で決まる。距離が共通変数であり存在の不安定性を起こし質点の運動を起こす。この論理は質量を持たない客観点にも当てはまる。予想点からの偏位があり得る。その偏位量は統計的である。偏位量により存在の信頼度が変わる。どの存在も完全ではない。存在の信頼度は偏位が殆どなければ存在信頼度は高いが完全ではない。小さな偏位での存在確率は比較的高い。存在の確率分布は偏位なしを中心とした正規分布である。実際に客観点が一カ所に安定して存在していることは、存在と持続に関してその位置で安定している可能性を意味している。その位置が最も適切な存在位置であり得るが少しは不適切であることもあり得る。物理的に存在する点には持続性があるがそれは可能性である。客観点が一点に存在していることは存在の適切性の高い位置ではあるがわずかな不適切性も含まれている。それは存在と持続の不適切性が少しはある自然の偶然性と安定化によって存在している。客観点の存在を含めて事象は統計的安定化過程に従う。自然の事象の本質を認識することでこの安定化に到達する。事象はその確率により変化する。それを安定化原理と名付ける。

2．存在性

(1) 客観的存在

空間内の客観点には位置と運動があると考える。点の位置とは他の点との相対関係であり、その運動はその相対関係の変化である。点の位置が変化すれば例えば原点との関係が変化する。上述の文は存在の概念的認識である。この認識を自然に幾らかの不安定性を伴う客観点の存在の実在的認識として用いる。点が概念的なものでなく実在的なものであれば参照枠内での対応点にいつも完全に一致するわけではない。それはその参照枠の対応点からずれる可能性がある。合致するとは概念であり実際には完全ではない。その精度は確率的である。一致とは決まっていない。その合致にはずれや偏位を伴う。予想される位置には普通存在確率分布の中央にて信頼性が高くなる。客観点がその中央にない場合はもっと信頼性の高い位置へ移動する。その位置は一般に中央部である。信頼性の相違は求心力を生じる、それが安定化である。空間が存在を規定する、時間が持続を規定する。客観的存在物はその存在空間の座標に照合されることによりばらつきや揺らぎが起こり、それゆえ散在性と揺動性が存在する。存在は時間がゆえに変動する。空間が散在性を起こし、時間が揺動性を起こす。

固形物、それは固形化されたエネルギー、即ち物体もやはり客観点の集合体と見なされる、よってその存在は参照枠内で表せる。実在空間内の実在の物体の存在は人の考える存在と一致するとは限らない。人の思考は物差しでありその量はスカラー量である。この物差しで測った量は統計量となる。散在性のある点の集合体はこの意味において物理的対象物即ち実在物である。実在の存在をその形や量を省略して幾何学的な一点で表す場合がしばしばある。エネルギーの存在はしばしばこの幾何学点に形や大きさを無視してその物理的エネルギー量を代表させる。

⑵ 存在の散在性

事象は存在する。存在は統計量であり統計上の不安定性を少し含んでいる、それは散在性である。客観点の存在確率分布上ぶれがないと予期している位置で信頼性は最高となりその周りが低くなる。この状態が存在の精度の上下に関与しており散在性と呼ぶ。

この状態は粒子線については簡単に理解し得る。電子や他の粒子のある瞬間の位置を考える時その位置は計算できる。しかしすべての粒子が計算した位置にあるわけではない。その値には統計的なばらつきがある、それが散在性である。

計算された位置は極大値で、その周りにも幾らか存在している。これが統計的存在性である。

⑶ 存在確率分布

存在性の精度は完全ではない。存在性には常に標準偏差（σ）で表される散在性がある。存在確率分布は期待値とその周囲の存在確率を示す。期待値からの偏位は推計的変数である。期待値は最頻値である。最頻値における信頼度は存在性の精度である。この信頼度が高いほど存在性はより安定である。標準偏差が散在性を規定する。標準偏差が小さいほどより安定な存在性である。その精度は標準偏差の逆数で決まる。単純に考えるために x 方向の分布について検討する。x 軸上の出現率はみな同じである。それゆえ確率分布は標準偏差（σ）の正規分布（Φ）で期待値（$x=a$）はその中央である。

C章で示した例の存在性は下記の式で表され、それを変動性に応用して処理する。この存在状態は安定化する次の２式で表される。次式は存在状態を表す。

(1)　$$\Phi = \exp[-(x-a)^2/(2\sigma^2)]/[\sigma(2\pi)^{1/2}]$$

σ は Φ の標準偏差　$\sigma > 0$

x は気温により変化するグラスの中の水の温度、そして a はこの存在の最適温度Ｘを最適温度（a）からの偏位温度とすると、$X=x-a$。式(1) は次式となる。

(2)　$$\Phi = \exp[-X^2/(2\sigma^2)]/[\sigma(2\pi)^{1/2}]$$

期待される温度（$X=0$）の信頼度は $1/[\sigma(2\pi)^{1/2}]$。この値が高いほど信頼は急激に上下するので信頼度が高いほど最適温度からの偏位は小さくなり、高精度な温度の存在である。標準偏差（σ）が上がるほど信頼性は低くなる。それゆえ（σ）は散在性を表しその逆数（1/σ）は存在の精度を表す。

⑷ 発現率と散在性

ある点におけるエネルギー量をＥとしPcを単位エネルギーの発現率とする。すると存在確率分布の標準偏差（σ）と単位エネルギーの発現率（Pc）は次の如くである。

$$(4) \qquad \sigma^2 = Pc(1-Pc)/E$$

Pcは1に比して充分小さいので、

$$(5) \qquad \sigma^2 \fallingdotseq Pc/E$$

発現率とエネルギー量の比は標準偏差の二乗と等しい。よってエネルギーの平方根と存在確率の信頼度は反比例する。また信頼度と発現率の平方根は反比例する。均等な発現率の空間内の客観点はどこでもすべて同じ存在確率分布の標準偏差を持つ。その空間内で安定であり動きはない。しかし存在確率の信頼度は100％ではなくその位置を変える可能性も共存する。

E．持続性とその揺動性

1．揺動性

ある一点への存在を考慮した時、ずれの可能性もあるがふらつきの可能性もある。存在性には基本的に二つの不安定要素がある。それらは散在性と揺動性である。存在には考えられている位置にあるかの散在性と呼ばれる不安定性だけでなく、質的な不安定性である持続運動の不安定性もある。これは揺動性と呼ばれる。持続性の高い安定した運動では殆ど揺動性はない。揺動性が高く簡単にふらつく運動をする存在や揺動性が低くふらつかない運動をする存在がある。この違いは揺動性の違いで表される。揺動性が低いとはその点が強くその位置に固定されている事を含めて高持続性を意味する。存在には一般に精度と恒度の二要素があり、それは一点上の客観点には偏位だけでなく揺動もあるということである。

2．持続性

事象は持続する。持続は統計量であり統計的不安定性があり、それが揺動性である。事象には持続性がある。持続は第二の量である。存在の持続も統計的である。急激な変化の確率は低いだろう。低速な変化の確率の方が比較的高い。無変化である持続の確率は高いが100％ではない。持続確率分布は無変化を中心とする正規分布である。

事象には存在性と持続性の二つの重要な要素がある。質点の例について質点には位置と運動がある。運動とは位置の持続性でもある。それは位置の距離値を時間の要素で微分して得られる。位置には存在確率分布があり、運動には持続確率分布がある。存在には散在性があり運動には揺動性がある。これら二つの要素は統計量である。事象には主要な二要素がある。それは存在と持続であり統計的な値として散在性と揺動性である。

客観点には二つの要素がある。それは位置と運動である。運動の不安定性分布は揺動性なしを中心とした正規分布である。その標準偏差は不安定性を示す。この持続確率分布の独立変数は存在確率分布の独立変数の時間微分値である。その変化率をここでは揺動と呼ぶ。静止した客観点は揺動極大値が0でありその信頼度が静止の確率である。確率分布は正規分布でありその標準偏差が小さいほど静止の可能性が高く持続性が強い。だから標準偏差の逆数を持続性と呼ぶ。客観点が一空間内の一点（a）に静止している場合は揺動速度が0である。静止状態でも揺動はある可能性がある。

3．持続確率分布

Xをx軸上の点（$x=a$）に留まっている場合の偏位量とすると揺動速度（V）は次式の如くである。

$$(6) \qquad V = dX/dt$$

揺動とは偏位の時間的変異率であり速度の次元である。客観点にはどの方向にも0を中心とする正規分布がある。（Ψ）を持続確率分布とするとその標準偏差（τ）は揺動性を表しその逆数は恒常性即ち持続の強さを表す。x軸上で考えてある客観点が点（$x=a$）上に静止していると揺動速度0を中心とした正規分布を持つ。

$$(3) \qquad \Psi = \exp[-V^2/(2\tau^2)]/[\tau(2\pi)^{1/2}]$$

$V=0$の信頼度は$1/[\tau(2\pi)^{1/2}]$であり固定の信頼度である。この値が高いほど少しのずれで信頼度は急激に下がる。

偏位（X）の点には持続確率がありそれは式（3）が示す正規分布（Ψ）である。持続性は完全ではない。持続性はいつも揺動性（τ）を伴う。持続性は完全ではない。

F．変動性と存続性

1．存在の安定性

存在性には持続性を伴う。それらは存続性や変動性そして実質性を形成する。客観点の存在などの事象には存続性と変動性がある。存在状況が変わると事象は安定化への変動を行い存続へと安定化する。事象には変動性と存続性があり、その結果として実質性がある。変動性（Z）については存在の確率（Φ）と持続の確率（Ψ）は比例する、そして存続の確率については存在の確率と持続の確率は反比例する。客観点の存在性には持続性を伴う。それらは変動性や存続性を形成しその事象を実在化する。客観点は実質性を持たねばならない。実質性においては変動性と存続性は反比例する。それゆえ散在性もしくは揺動性が0になれば実質性もなくなる。実在物が実質性を失えばそれは宙に消えてしまう。実在物は実質性を持たねばならない。散在性か揺動性が0になれば実質性は空間から消失する。事象は変動性と存続性のバランスに基づいて存続する。それが実質性の特性である。

後で述べるように存在性及び持続性の信頼性が極大の場合、変動性は式（7）が示す如く揺動性と散在性は比例し、存続性においては式（8）が示す如く揺動性と存在性は反比例する。変動性と存続性は式（9）の示す如く反比例して実質性をなす。すると散在性か揺動性が0になれば実質性は消える。存在物が実質性を失えばそれは宇宙に消える。存在物は変動性と存続性のバランスの基に存在する、それが実質性の安定した特性である。

2．客観点の変動性

変動性においては散在性と揺動性は同じ方向にある。変動性では散在性と揺動性が比例する。散在性と揺動性が比例するゆえ、変動性を持続確率と存在確率の比と定義する。後で述べるが変動性（Z）は定数であり時間経過には関係しない値である。概念点は概念空間内で自由に動くことができる。実在点は空間内で自由には動けない。特に一形態上の各点の実在空間原点に対する相対関係に関して制限がある。実質性の即ち実際の運動は持続確率だけでなく存在確率にも関わっている。運動は時間に関係する。その上存在確率が動きを制限する。

運動を考えるに当たっては時間経過の概念が必要である。時間経過をここでは時間と呼ぶがある瞬間から他の瞬間までの経過である。運動は存在信頼度を変化させて客観点を安定化する、それが変動性である。変動性には時間が含まれる。揺動性の高い客観点は変化し易い、散在性が高いともっと変化し易い。変動性では散在性と揺動性は同じ方向である。散在性と揺動性は変動性においては比例する。変動性は式（7）が示すように存在確率と持続確率の比であると定義する。後に記述するが変動性（Z）は時間経過に関係しない定数である。

(7) $$Z = \Phi/\Psi$$

式（7）の両辺を対数化して微分する。

(8) $$(d\Phi/dt)/\Phi - (d\Psi/dt)/\Psi = (dZ/dt)/Z$$

変動性は存在性の変化率と持続性の変化率の差が変動性変化率に等しくなる。上式に式（2）と式（3）を代入すると下式を得る。

(9) $$(dX/dt)X/\sigma^2 - (dV/dt)V/\tau^2 = (dZ/dt)/Z$$

X は最適値（a）からの偏位、そして $V = dX/dt$。Z は定数。

すると下式を得る。

(10)　$d^2X/dt^2/\tau^2 - X/\sigma^2 = 0$　　$\tau/\sigma = \omega$とすると、

(11)　$d^2X/dt^2 - \omega^2 X = 0$

式（11）は下式のように変化できる。

(12)　$[d/dt + (\tau/\sigma)][d/dt - (\tau/\sigma)]X = 0$

(13)　$dX/dt - (\tau/\sigma)X = 0$　　または、

(14)　$dX/dt + (\tau/\sigma)X = 0$

式（13）は無限に拡散するのでここでは現実的ではない、これは安定化過程であり式（12）の解として下式が適応される。

(14)　$dX/dt + (\tau/\sigma)X = 0$

この式に $\tau/\sigma = a$、$X = x - a$ とすればC章で述べた $dx/x = -b*dt$ と同じになる。上式（14）の解は次式となる。

(15)　$X = X_0*\exp[-(\tau/\sigma)t] + C$

Cは定数

現実には存続性も作用するゆえ、この事象の変化はこのカーブではなく減衰波動関数となる。

３．客観点の存続性

存続性に関しては散在性の高い事象には揺動性が低くなければならないし、揺動性の高い事象には散在性が低くなければならない。安定な実質性は変動性と存続性から成る。自然の事象は継続するか変化するかである。変化は本質的には安定性即ち存続性への変化である。状況が変化すれば存続性への変化をする。事象の存続性では高い散在性には低い揺動性にならなければならない。また高い揺動性には低い散在性にならなければならない。即ち存続性においては散在性と揺動性は逆方向の存在である。存続性では散在性と揺動性は反比例する。存在信頼度と持続信頼度は反比例している。存続性は式（16）が示す如く存在確率と持続確率の積である。後に述べるが存続性（H）は定数である。

（16）　$H = \Phi * \Psi$

４．実質性

事象が変動しつつある時は変動性が大きく存続性が小さい。そしてそれが安定な時はその変動性は小さく存続性が大きい。変動性と存続性は反対の方向である。それらは物理的実質性においては反比例している。存続性の低い事象には高変動性があり簡単に変化する。しかし低変動性の状況においては高存続性を持つ。安定した高存続性の事象は低変動性である。だから実質性（Θ）は式（17）が示す如く変動性（Z）と存続性（H）の積である。実質性（Θ）は定数であり事象の特性と本質を表している。

（17）　$\Theta = Z * H$

事象の実質性については式（17）が示す如く変動性と揺動性は反比例しているのであるが、それが変動性や存続性が定数であることを示している。実質性は事象や存在の特性を表しているので定数である。それで変動性も揺動性も定数であることを示されている。

式（17）の両項の対数を取る。

$$\log(Z)+\log(H) = \log(\Theta)$$

そして両項とも微分する。実質性は事象の特性であるから定数である。

だから $d\Theta/dt/\Theta = 0$。よって、

$$(18) \qquad dZ/dt/Z+dH/dt/H = 0$$

周囲の状況が変わらなければ存続性も定数、結果として $dZ/dt/Z$ も０。状況が変わらなければH（エータ）もZ（ゼータ）も定数である。

存在は存続性があるが変動性もあるのは存在の散在性と揺動性に基づく。変動においては新たな実質性を生じるので存続性や変動性を新たにした存在を生じ得る。

G．回転と安定化

1．事象の回転

(1) 変動性と回転

変動性においては存在性の信頼度と持続性の信頼度は比例する。だから変動性は存在性確率（Φ）と持続性確率（Ψ）の比である。次式に示すように変動性は比例定数となる。

$$(7)\qquad Z = \Phi/\Psi$$

両項の対数を取り微分すると次式を得る。

$$(19)\qquad (d\Phi/dt)/\Phi-(d\Psi/dt)/\Psi = (dZ/dt)/Z$$

存在性の変化率と持続性の変化率の差が変動性変化率となる。式（2）と（3）を上式に代入すると下式を得る。

$$(20)\qquad (dX/dt)X/\sigma^2-(dV/dt)V/\tau^2 = (dZ/dt)/Z$$

X は存在点（a）からの偏位であり $V = dX/dt$ で Z は定数、すると下式を得る。

$$(21)\qquad d^2X/dt^2/\tau^2-X/\sigma^2 = 0$$

τ/σ の次元は角速度の T^{-1} であるから $\tau/\sigma = \omega$ とおくと下式を得る。

$$(22)\qquad d^2X/dt^2 = \omega^2X$$

客観点 $x = a$ は回転して正の加速度が存在する、それが遠心力を起こしている。

⑵ 存続性と回転

存続性とは安定した存在である。存続性（H）では存在性確率（Φ）と持続確率性（Ψ）が反比例しており、それらの積である存続性（H）はそれらの反比例定数である。客観点（A）が x 軸上の位置（a）に静止している場合の存続性は、

$$(16)\qquad H = \Phi * \Psi$$

上式の両項の対数を取り微分する。

$$(23)\qquad (d\Phi/dt)/\Phi + (d\Psi/dt)/\Psi = dH/dt/H$$

上式に式（2）と（3）を代入すると下式を得る。

$$(24)\qquad (dX/dt)X/\sigma^2 + (dV/dt)V/\tau^2 = dH/dt/H$$

X は x 軸上（a）点からの偏位、V は V = dX/dt、H は定数、よって上式は下式となる。

$$(25)\qquad d^2X/dt^2/\tau^2 + X/\sigma^2 = 0$$

τ/σ の比は回転角速度（ω）である。上式は波動関数で、客観点は点（a）上で回転している。よって上式は下式となる。

$$(26)\qquad d^2X/dt^2 = -\omega^2 X$$

存続性は波動である。存続性は波打っている。点（a）の位置にある客観点は回転していて負の加速度を持ち回転による求心力を起こす。式（22）による遠心力と式（26）による求心力により客観点は安定である。安定した客観点には角速度があり点（a）の位置で回転している。

式（22）は点（a）上の客観点が回転して陽性の加速度を有し遠心力を起こすことを示す。そして式（26）がその客観点が陰性加速度を有することを示す、それは回転求心力を起こす。双方の力でその点の安定を

保つ。回転事象が安定した実在点を実現する。実在点は回転しているから概念点と違い次元があると考えられる。

⑶ 実質性とエネルギー

物理的な事象は通常スカラー値で認識される。しかしその値は変動性と存続性のバランスの基に定まる統計的な最頻値と考えるべきである。

客観点 $x=a$ は変動性により回転して安定化している。存在の散在性が大変高い場合は存在性確率分布が平坦に近くなる。そして同じ実質性であるから揺動性も大きい。そうすると回転は波動となる。そしてその波動は変動性により正の方向であり、また存続性による波動は逆方向である。これらの波動はエネルギー波である。双方向の波動の力は（τ/σ）で決まるので同じである。実質性（Θ）は事象の特性であり本質である。実質性はエネルギーの力を示す定数である。後に述べるがエネルギーは実質性定数に基づくだけでなくその散在性や揺動性にもよる。エネルギーの大きさは波動源の量の大きさに基づく波高による。

2．空間の安定化

空間は参照枠の中で規定され存在する。この空間内の客観点には質量はないが、実質性であり参照枠内の対応点に対して種々の偏位のある位置や種々の変動のある運動がある。実在空間の時間は存在を変化させる。その空間内の点はいくらかの不安定性があるゆえ存在確率や持続確率があり、それらが存続性や変動性をもたらす。それゆえその点の質量はないが実質性がある。実在空間は主観的な概念点と共にこの種の客観点が存在するゆえ実質性であり不安定性もある。実在点の変動性は式(22)に基づく遠心加速度を起こす。存続性は式（26）に基づく求心加速度を起こす。両式による客観点の運動は安定な微小空間の存在をもたらし、その中の客観点は安定に回転している。このような実在点の微小空間で構成される空間を実在空間と呼ぶ。実在空間は安定である。

3．点運動の安定化

もし x 軸上の位置（b）に客観点がありその存在確率が位置（a）のそれより低いと位置（b）上の点は x 軸上の位置（a）に向かって動き出す。位置（b）は位置（a）の偏位と見なされる。位置（a）からの求心力位置（b）の客観点に作用する。その客観点が x 軸に沿って動いているなら上記式（10）は次のように変化できる。

$$(12)\qquad [d/dt+(\tau/\sigma)][d/dt-(\tau/\sigma)]X=0$$

$$(13)\qquad dX/dt-(\tau/\sigma)X=0$$

解として式（13）は無限に拡散する破壊過程であるから現実的ではない。これは安定化過程でありここでは収束過程の次式が適応されるべきである。式（12）の解は次の如くである。

$$(14)\qquad dX/dt+(\tau/\sigma)X=0$$

この式の解は下記の如く減衰関数で X_0 は X の位置（b）における初期偏位である。

$$(15)\qquad X=X_0*\exp[-(\tau/\sigma)t]+C$$

Cは定数

偏位は減少係数（τ/σ）で減少し 0 に収束する。それは式（26）の波高が 0 に収束し点（b）の存在信頼度がなくなり点（a）の信頼度が上昇する。しかし存続性が作用するのでこの解は現実的ではない。客観点は式（14）の示す変動性により運動し式（25）が示す存続性で安定化する。実質性によりこれら２式の和が運動を表す。式（14）が変動性による（b）の位置から（a）への運動を示しており、式（14）は次のように変更できる。

(27)　　$2\omega dX/dt+2\omega^2X=0$

存続性も運動に寄与しており式（26）を次のように変更する。

(28)　　$d^2X/dt^2+\omega^2X=0$

式（18）によれば運動の実質性は式（27）と（28）の和であるから下式となる。

(29)　　$d^2X/dt^2+2\omega dX/dt+3\omega^2X=0$

上式の解は次の如くである。

(30)　　$X=\exp(-\omega t)*[X_1*\cos(2^{1/2}\omega t)+iX_2*\sin(2^{1/2}\omega t)]$

X は実数でなければならないから $X_2=0$。
$t=0$ の時は、X は初期偏位 X_0、よって $X_1=X_0$
すると式（31）が得られる、それは減衰波動関数である。

(31)　　$X=X_0\exp(-\omega t)\cos(2^{1/2}\omega t)$

波高 X は減少し 0 に近づいて客観点は安定化する。客観点の安定化は質点における引力の法則の如くこの過程に従っている。客観点の運動に関しては式（31）は（τ/σ）を定数としているので近似式である。現実には存在性は上昇し持続性は下がる、即ち（σ）は下がり（τ）は上がる。点融合においては存在性や持続性は変化する。上述の過程は客観点の融合過程の概略を説明している。この種の融合点が巨大な塊である場合は式（22）と（26）は物体の質点を指しておりドブロイの物質波を意味している。エネルギーは大変高周波の空間波あるいは物質波である。

H．融合した客観点

1．客観点の融合

客観点には位置と運動があるから質点のように散在性（σ）と揺動性（τ）がある。実際に客観点が一点に安定してとどまっていることはその客観点は統計的に最も適切な存在であるがわずかな不適正もある。一点上の客観点には散在性があるがそれは点融合で殆ど変化しない、またその運動には揺動性があるがそれは融合に当たって上昇し持続性を減少させる。ある空間内には無限数の主観点（ユークリッド点）があると同時に大多数の客観点がありそれらには次元はない。これらの客観点にはいろいろな実質性があるが、ただの点であり次元も持たないのでその存在性はおよそ均一であると考える。しかしそれらの持続性は不均一である。この状態は客観点の融合によって発生する。次のような過程が考えられる。多数の点の融合は存在性（Φ）と持続性（Ψ）の比である変動性（Z＝Φ/Ψ）を高めることが期待される。点融合に際して存在性は殆ど変化しない。持続性は減少する、それは揺動性（τ）が上昇する事を意味する。客観点は融合する、多重の融合は持続性を減少させる、そのことは変動性を上昇させ式（22）と（26）が示すように、より高周波のエネルギーを放出する。τ/σ の比は回転角速度（ω）である（τ/σ＝ω）。ω の次元は T^{-1}、それは角速度の次元である。式（22）は変動性の波動関数で陽性の方向に放出されるエネルギー波を示す。そして式（26）は存続性の波動関数で陰性の方向に放出されるエネルギー波を示す。それゆえ融合した客観点は安定でありその位置を維持する。

客観点融合はポアソン確率分布に基づいて発生するから高次な点融合は少ない。高次な融合発生頻度は低い。大多数は低回数の融合点である。融合点の融解も起こり得る、それによってはより低いエネルギーも発生する。

2．集合体

空間においてはいろいろ客観点融合が起こる。客観点は各種多数の融合客観点の集合体を形成する。それらはなお一点で代表される存在であり、それらは固有の実質性を有する。その発生はポアソン確率分布によるゆえ低レベル集合体の発生が高レベルのものより大変高くなる。集合レベルの相違は同じ融合数であっても実質性の違いが発生する。即ち客観点の融合数の違いはなくても実質性の相違は発生する。集合体には固有の実質性がある。弱いエネルギー波を発生する大変低い集合体は感知できない。それらを非触エネルギーと称する。最も弱い可触エネルギーは熱である。熱以上に強いエネルギーは可触エネルギーと称する。

3．微　泡

客観点には次元がないゆえ、その集合体にも次元はなく１点上での存在である。その１点上での実質性を有する。大きな集合体は実質性が高く回転等運動も強く、その存在にはいくらかの空間が必要である。いろいろな集合体の構成はそれ相応の微小空間が必要と考え得る。それらをここでは微泡と名付ける。１微泡内に複数の集合体の存在も考え得る。それがエネルギー発生を考える最小の空間である。微泡内では単独ないしいろいろなレベルでの実質性のある客観点の集合体が発生する。発生する集合体のレベルはポアソン確率分布による。大多数は非触エネルギーのレベルである。だから微泡内での可触エネルギーレベルの集合体では熱エネルギーレベルが最高である。最小可触レベル未満の集合体発生はここでは省略する。だから微泡内容を代表するのは最小可触エネルギー即ち熱レベルである。微泡内では可触エネルギーで最大量なのは０

ケルビンの熱エネルギーとし、それが微泡の実質性を代表する。熱エネルギー以上の可触エネルギーは多くない。しかしもっと高いエネルギーレベルの実質性も微泡内に存在し得る。

4．非触エネルギー

実質性を伴う客観点の偶発的な結合が起こり得る。それらが集合体である。それらには偶発的な存在性と持続性の違いが起こり、より高い実質性を誘発するがなお１点である。その１点の運動には微小空間を必要とする。集合体の実質性が回転を起こしエネルギー波を発生する。エネルギー波の振動数が０ケルビン未満のものは非触エネルギーである。集合体が融合する統計的な過程において実質性が上昇することにより変動性が上昇する、それがより高いエネルギー波を発生する。微泡内で融合したいろいろな集合体はいろいろな高振動数のエネルギー波を放出する。微泡内の集合体の殆どは非触レベルの集合体である。エネルギーには質と量がある。量は波高即ち微泡等波源の量によるが質は振動数による。０ケルビン未満のエネルギー波は感知が困難であるがいろいろな事象を修飾することができる。例としては音や波のように簡単に空気や水の存在を修飾する。度重なる集合体の融合による高い揺動性から高エネルギー周波は発生する。集合体融合の頻度はポアソン分布による。だから空間内のエネルギー波の多くは感知不能で熱エネルギー未満である。波の力即ち波の振動数は実質性の上昇による、それは集合体融合の上昇であり微泡の併合により促進される。

5．微泡の併合

一定の空間内に一定数の同じような基礎的構成の集合体群を有する微泡の存在は考え得る。微泡には多量の集合体があり可触エネルギーレベルの存在確率では熱レベルのものが最も高くそれ以上は減少する。各々の微泡内には殆ど同じ基礎的構成の集合体があると考える。そのエネルギーレベルは最弱可触エネルギーでは熱エネルギーで代表される。空間とはそのような微泡の集合体と考えられる。微泡は平均的に殆ど同様な構成の集合体を有していると考える。微泡は平均して類似の実質性の集合体を有する。基礎的微泡が度々併合して種々の集合体をポアソン分布に従い形成する。併合した微泡内では集合体は互いに融合し融合回数により高い実質性の集合体を得る。より高い融合回数はより高い実質性にしてより高いエネルギー波を発生する。度々融合したより大きな集合体はより強いエネルギーを発生する。エネルギーの強さは実質性の周波数による、またエネルギーの量は同レベルの集合体の量による。

Ⅰ．可触エネルギー

1．微泡併合と高エネルギー

微泡併合がもっと強い実質性をもたらす。高頻度の微泡併合は高頻度の集合体融合をもたらし、それにより高実質性を発生し高周波の可触エネルギーをもたらす。それらはいろいろな存在性や持続性の確率を持ち各種可触エネルギー波を発生する。

0ケルビンのエネルギー波を発生する実質性（Θ_1）の集合体を基本とする微泡を標準とする。するとその（λ）倍の実質性（$\lambda*\Theta_1$）である集合体を有する微泡も考えられ、それらの微泡併合は実質性を顕著に高め高エネルギーを発生する。空間内一定の時間内に起こる期待微泡併合回数（λ）を検討する。P(λ) は併合微泡の存在確率であり下記の等式に基づき検討する。

微泡併合は実質性を比例的に増加する結果高エネルギーを発生する。それら微泡のポアソン分布による併合確率は対応する集合体をもたらす。所定時間内で所定空間内での予想実質性増加（λ）の出現確率を検討する。k は各λ値についての適正確率（極大確率値）を定めるための仮想併合回数である。ここでは微泡併合の増加と集合体の増加は比例するとする。P は併合微泡即ち融合実質性の出現確率、λ は微泡併合の期待回数で、k は各λ値における P の適正度を定めるための仮想微泡併合回数であり下記のポアソンの式に基づく。

(32)　　$$P(k, \lambda) = \lambda^k * \exp(-\lambda)/k!$$

λ：微泡併合の期待回数

P：併合微泡の出現率

k：仮想微泡併合回数

客観点には実質性があり波動を発生する。波動には力と力量がある。力は波数に依る故λの最適値による。力量は同じ程度の実質性量が規定する波高に依る。

2．エネルギー発生

微泡の多重併合は多数の基準集合体の融合を起こし高度な実質性（Θ）をもたらす。高度な実質性は基準実質性（Θ_1）が度々（λ）回融合した結果強いエネルギーを発生すると考える。高エネルギーの存在確率（P）は融合回数（λ）が増えるに従い上昇する。すると（P）はピークに来てその後は減少する。また（λ）が大変小さい時はそのピークは殆ど（λ）の始まりの所に起こる。λが1の時ピークは最初、又はその近くにあり急速に低下するからこのような小さな値は現実には存在しないと考える。（λ）の値が大変上昇するとPの存在確度分布は正規分布に近くなる。各（λ）についてkの値によるPの値にはピークがありそのピーク時のkの値即ちkの適正値はkの平均値であり、その平均値はポアソン分布ではk＝λであるから、（λ）回融合した実質性の存在確率Pはk＝λの場合であり、下式で表される。

$$P(k, \lambda) = \lambda^{\lambda}/\lambda!*\exp(-\lambda) \quad (33)$$

客観点は波動をなす実質性である。波動には強度と力量がある。強度は実質性（Θ）の強度即ち振動数による。（Θ）は（$\lambda*\Theta_1$）である。融合の適正回数は上式の極大値を成すλなので、そのλが強度を定める。そしてその力量は同様な実質性の量によって決まる。

3．可触エネルギーの種類

実質性は存在性と持続性から成る。よって実質性は散在性と揺動性による。実質性が同レベルであってもエネルギーの種類は散在性や揺動性の値によっては違った種類となる。だから可触エネルギーは下記の通り分類できる。

A．高散在性の点が発する波動エネルギー

1．低揺動性とでは熱の如く低振動数の波動
2．高揺動性とでは光の如く高振動数の波動

B．低散在性の点が発する粒子エネルギー、質量があるものないもの

3．低揺動性とでは量子の如く低回転数の粒子エネルギー
4．高揺動性とでは質量の如く高回転数の粒子エネルギー

4．熱

高散在性の熱エネルギーは最小の可触エネルギーである。本来集合体がその実質性により回転して＋と－の二方向にエネルギー波を発生する。集合体群は一定方向を向いているわけではなくあらゆる方向にエネルギー波を発生させている、散乱性の強いエネルギー波であるから空間に充満するエネルギーである。

5．光

光は波である。光は周波数が大変高い直進する客観点の波である。成因は熱と殆ど同じであるが散在性が少ないため空間内をあらゆる方向に直進する。赤色光は毎秒約4500億サイクルである。そして青色光は毎秒5500億サイクル。一般の可視光線は毎秒4300億から7700億サイクルである。これらは高散在性高揺動性の実質性が発生する高周波である。光には粒子の性格もあり、それが量子である。

6. 量　子

空間内多数の実在点から発生する高周波は均一ではない。いろいろな光の高周波が重なり合うとその波の波長は不規則になる。低周波の部分や高周波の部分が発生する。極端に高い高周波の部分が固形化し量子となる。

7. クォーク

低散在性の客観点の発生するエネルギー波は直進性が弱く自転する。低散在性で低揺動性の客観点はクォーク、中間子等の質量のない粒子となる。

8. 質　量

低散在性で高揺動性の客観点は質量を持つ。

J．実質性とエネルギー

1．実質性の力と弾み

実質性には存在性がありそれには少々不安定性がある。存在性にはその位置的状況を正常に戻す確率、即ち能力がある。それは式（34）の示す力である。位置上で力を積分したものが存在性である。位置上で積分した力はエネルギーでもある。

（34）　　$dE = F*dx$

よってエネルギーの変動は力に存在性の変動を乗じたものと定義できる。

持続確率の独立変数は速度である、この変数は少しのふらつきがある。それは時間的な不安定性であり加速度を成す。持続性は正常な持続性を成すための加速度を起こし得る。それは実質性からの力である。弾み（P）の変化は力を発生する。力と時間変動の積が弾み変動となることを式（35）が示す。

（35）　　$dP = F*dt$

高い変動性は低下する傾向があり、より強い力を発生し、その時間的集積は弾みとなる。それはエネルギーである熱として移動するだろう。また、それはエネルギーである光のように真っ直ぐ飛んでゆくだろう。するとその客観点の変動性は低くなる。温度はエネルギーの一種であるが実質性の変動性を表している。温度が低下すると実質性は安定化する。それは実質性が存続性を上昇させ変動性を低下させるゆえ弾みは減少する。式（35）から弾みの変動と時間変動の商は力である。式（34）

のFに dP/dt を代入する。

$$dE = dP/dt*dx \quad (36)$$

位置変動と時間変動の商は速度である。エネルギーの変動それは存在性の変動であるが弾み変動と速度を乗じたものとなる。

$$dE = dP*v \quad (37)$$

2．実質性と弾み

実質性には存在性と持続性がある。安定した空間は大多数の同一実質性の客観点から成り、バランスも良く動きもない。安定性は空間内で実質性の均一分布による。

偶然には空間内に実質性の異なった客観点が存在し得る、またその位置的に合わない違った実質性値の客観点も起こり得る。それらは安定化のため動き出す傾向がある。しかしこの実質性はこの客観点の周囲全方向に同じ力の実質性であるからこの客観点は不動である。近くにもう一つ相違する実質性の客観点がある場合は双方の弾みが影響し合って互いの方向に移動する、その場合両方の客観点の実質性の変動性は減少する。それらの客観点は融合し新たな実質性を得る。このようにして客観点は周りの点と反応して空間内での位置的な実質性のバランスを得るのである。空間内での不均一分布は客観点の実質性の移動を起こし変動性が変化する。実質性の移動は弾みである。不均一な実質性分布はその移動や弾みの変動を起こし、それにより存在性変動が誘発される。実質性は定数ではない。実質性の変化は弾み変動を誘発する。移動は式（37）が示す如く存在性変動や弾み変動に影響する。

3．実質性の固形化

客観点の実質性には変動性と存続性がある。変動性は回転し遠心力を誘発し存続性は回転して求心力を誘発する。客観点は回転していて単なる一点ではない。その回転は微小空間を取り得る。変動性が極めて高い実質性の客観点は極めて高回転数での遠心力がある。またそれらで極めて高い存続性を有する客観点は極めて高回転数での求心力がある。そのグループの客観点は互いにくっつきあい固形化する。それが質量又は物質である。

それら固形化した客観点は互いに引き締め安定した変動性と存続性を得る。そのような固形化した客観点は質量であるがそれ固有の高存続性、低変動性の大変安定した実質性を得る。

4．質　量

質量をここでは固形化した実質性と同等と見なす、即ちその存続性は大変高いが少しは変動性もある。だから質量は固形の存在でその存続性は非常に高くその変動性は非常に低いにもかかわらず、いつもその条件に基づいていろいろな運動に関与している。質量はただの客観点の集まりとは違った存在である。質量はそれ特有の変動性や存続性や実質性を獲得している。その実質性には大変低い変動性と高い存続性がある。この存在物は高振動の実質性でその回転頻度が極めて高く前進性でない客観点の集まりである。電子の物質波の波長は可視光線の波長の1000分の1である。それらは回転客観点の集合体である。

5．修飾されたエネルギー

ある種のエネルギーが他種のエネルギーの現状を修飾することがあ

る。例としてお湯である。熱エネルギーが固形化されたエネルギーの水の実質性を修飾する。強力な熱エネルギーが液体状態の水を過度に熱して蒸気の実質性を発生する。それは新しい存在物即ち新しい実質性のエネルギーの形態である。

実質性の運動は弾みやエネルギーを変化させる。エネルギーのこの変化は実質性の状態を変化させる。実質性はその形態を変化する。この場合の実質性は水の変動性を上昇させ存続性を低下させる。不安定状態のエネルギー波に上昇した変動性と低下した存続性が存在する。実質性はその形態を変えることである。

６．質量とその修飾運動

質量は固形化された客観点の集合体でありそれ自体の大変安定した実質性であるが、いろいろなエネルギーの要素に修飾される。客観点の量即ち結合した実質性の大きさは一定の実質性で質量（m）として表されるが、その流れ（v）はエネルギーの時間変化（$\mathrm{d}E/\mathrm{d}t$）である弾みである。質量の動きは弾みである慣性となりその積分がエネルギーである。

質量の単純な定速運動は慣性である。

$$P = m*v$$

その積分はエネルギー、

$$E = m*v^2/2$$

力は質量と加速度の積である。

$$F = m*\mathrm{d}v/\mathrm{d}t$$

7．波（津波）

津波の実質性は変動性が低く存続性が大変高い水の波である。それは多量の水の実質性が大変強力な低周波の非触性のエネルギーである波に修飾されたものである。その動きは水の波エネルギーを大変強力にしている弾みである。その弾みの強化は水の実質性によるものではなく波エネルギーの強さによるものである。水の実態は高存続性で低変動性の実質性である。この実質性の流れの弾みが波のレベルまで強化されたのである。

8．エネルギー保存の法則

式（37）は運動のない事象には存在性の変化がないことを示している。存在性はエネルギーに比例するのでエネルギーは保存される。

$$(37) \qquad dE = dP{*}v$$

実質性は変動性と存続性から成り弾みを発生する。各客観点にはそれ固有の弾みがあり安定している。ある事象が起こりその空間内の客観点はその周囲の点と反応して実質性を増加また減少する。その事象の前後で各客観点の実質性の値は違っているだろう。その事象が安定すると弾みの値の合計は実質性の流失がなければ不変である。

$$(35) \qquad dP = F{*}dt$$

式（35）は弾みに変化がなければ力はないことを示している。式（37）は運動や弾みに変動がない時はエネルギーに変化がないことを示している。エネルギー保存の法則はこれらの状況に当てはまる。式（37）は弾みの合計に変化がなければ存在性の合計にも変化がないことを示している。存在性はエネルギーと比例している。よって存在性の不変はエネルギーの不変でありエネルギーは保存されている。

K．仮想的存在性とエネルギー

1．仮想的存在

散在性が負である仮想的存在性は考え得る。そのエネルギーがその存在を表している。そして変動性が負である仮想的状態でのエネルギー過剰やエネルギー欠損の客観点を考え得る。

式（2）における標準偏差は散在性を明示している。標準偏差が増加するに従って客観点の信頼性は低下する、そして散在性は無限になればその点の実質性はなくなる。負の散在性を用いて仮想点に式（2）を応用する。X は仮想点からの距離であり、そこのエネルギーは過剰の場合も欠損の場合もある。

(2) $$\Phi = \exp[-X^2/(2\sigma^2)]/[\sigma(2\pi)^{1/2}]$$

エネルギー過剰の場合は遠心力となる。そして標準偏差の絶対値が小さいほど遠心力は強くなる。過剰分（E）は減少する傾向がある。エネルギーの欠損ではエネルギーを引き寄せ欠損（$-$E）を修復する傾向がある。

力はエネルギーと Φ の積であるから遠心力は（$1/\sigma$）に比例する。式(2) から中心力（F）は［$1/\sigma(2\pi)^{1/2}$］に比例する。

(38) $$F = E/[\sigma(2\pi)^{1/2}]$$

式（38）を応用すると比例定数の E は過剰エネルギーと考えられ F は負で遠心力である。式（38）は散在性が無限大になれば遠心力は消失することを示している。散在性が陰性もしくは陽性の無限大の時の存在信頼度の平坦な分布になる。エネルギーの存在確率分布が平坦な濃度になることは式（38）の E が示している。

2．力の実体化

期待位置（X＝0）でのエネルギー過剰のもろさは$1/[(2\pi)^{1/2}\sigma]$である。この値の絶対値が高いほど過剰エネルギーはよりもろくなるので、エネルギー過剰は圧力を受けより平らになる。もろさは標準偏差の絶対値が下がるほど増加するのでその逆数（1/σ）が低下する圧力を示す。遠心力はその圧力（1/σ）に比例する。式（2）から中心力（F）は$[1/\sigma(2\pi)^{1/2}]$に比例する。式（38）に示す比例定数（E）は過剰エネルギーと考える。このEが単位エネルギーである場合はその確率分布は偏位X上のエネルギー分布となる。式（38）において−Fは分布中心の圧力である。

3．実質性の仮想的実態

素粒子はエネルギーのある実質性存在物である。実質性存在物である中性子が一例である。中性子はエネルギーから成る陽子と電子に成り得る。中性子が電子を放出する時新しい存在物である陽子が完全にバランスの取れたエネルギーを得ることができない。それは一時的状態であるがエネルギー過剰が新しい陽子に残され得る。同時に電子にも一時的状態であるエネルギー不足が起こる。これら陽子と電子はクーロン力を起こし、ある角速度で陽子の周りを旋回する電子が存在するのである。高周波のエネルギーである強力なエネルギー波は量子化や粒子化する。粒子はそれ特有のエネルギーレベル（中性子）がありエネルギー過剰（陽子）やエネルギー不足（電子）が起こる。

4．クーロン力

+クーロンと考えるエネルギーの小山Eはもろくその近辺に遠心力

を発生する。エネルギーEの中心部のエネルギー影響量は $E/[\sigma(2\pi)^{1/2}]$ でその遠心力は式（38）を用いて求められる。エネルギーEの存在確率分布上偏位Xの位置にエネルギーの谷（−K）があるとエネルギーEの存在状況はエネルギー（−K）の存在確率分布から影響を受ける。エネルギー（−K）が遠心力のエネルギーEを変化させる。エネルギーEの中心にかかる力はエネルギーEの存在性とエネルギーEの中心におけるエネルギー（−K）の存在性との積である。この場合σは共通でFは陰性の求心力となる。この求心力は次の式で求められる。

(39)
$$F = E\exp[-0^2/(2\sigma^2)]/[\sigma(2\pi)^{1/2}]$$
$$*(-K)\exp[-(-X)^2/(2\sigma^2)]/[\sigma(2\pi)^{1/2}]$$
$$= (-K)E\exp[-X^2/(2\sigma^2)]/(2\pi\sigma^2)$$

Xが十分大きければテイラー展開を用いて近似される。

$$= (-K)E[2\sigma^2/(-X)^2]/(2\pi\sigma^2)$$
$$= (-K)E/(\pi X^2)$$

上式は求心力であるから、

$$-F = KE/(\pi X^2)$$
$$-F = E/(\pi^{1/2}X)*(-K)/(\pi^{1/2}X)$$

陽性クーロン（Q）と陰性クーロン（−Q）の積が求心力と定義されている。

$$-F = Q(-Q)/X^2$$
$$-F = Q/X*(-Q)/X$$

エネルギーＥとクーロンＱは次式（40）の関係となる。

$$E = Q\pi^{1/2} \qquad (40)$$

エネルギー量とクーロン量の次元は同一である。

エピローグ

すべての存在物には存続性がある。しかし永遠に存続する実在物はないと言える。必ず変動性を大なり小なり伴う。それが実質性である。すべての実在物には実質性がある。安定化原理に基づけば実質性はエネルギーである。すべての存在事象はエネルギーから成り立っていると言える。変動性の高いエネルギーも存続性の高いエネルギーもあり得る。実在する事象はエネルギーによる存在であり、その変動はエネルギーによる変化である。事象の存在はエネルギーの存在である。そのエネルギーが変化するか、あるいは外からのエネルギーで変化させられることにより事象の存在は変化する。

赤沼　篤夫（あかぬま　あつお）

1964年３月東京大学医学部医学科を卒業後、放射線医学教室に入り放射線の研究を始める。1968～1973年まで米国とドイツに留学し、エッセン大学助教やストーニイブルック大学医学部講師を兼務。ブルックヘブン国立研究所で客員研究員を併任し、粒子線の研究を行う。1973年帰国し東京大学講師。1978年には米国ロスアラモス国立研究所でパイオンの医学利用の研究に携わる。1985年東京大学准教授に。重粒子線の研究のため1991年放射線医学総合研究所に転勤したが研究が進まず、独自に研究するため1997年退官。現在に至る。

【著書】

『安定性理論　The Theory of Stability』（2018年、東京図書出版）
『安定化原理　Stabilization Principle』（2020年、東京図書出版）

安定化原理に基づくエネルギー発生

Energy Development on the Stabilization Principle

2023年２月７日　初版第１刷発行

著　　者　赤沼篤夫
発 行 者　中田典昭
発 行 所　東京図書出版
発行発売　株式会社 リフレ出版
〒112-0001　東京都文京区白山 5-4-1-2F
電話（03）6772-7906　FAX 0120-41-8080
印　　刷　株式会社 ブレイン

ISBN978-4-86641-595-6 C3042
Printed in Japan 2023